FACADISM

FACADISM

Jonathan Richards

Routledge
Taylor & Francis Group
LONDON AND NEW YORK

First published 1994
by Routledge
2 Park Square, Milton Park, Abingdon, Oxfordshire OX14 4RN

Simultaneously published in the USA and Canada
by Routledge
711 Third Avenue, New York, NY 10017

Routledge is an imprint of the Taylor and Francis Group, an informa business

© 1994 Jonathan Richards

Typeset in Scantext September by Solidus (Bristol) Limited

All rights reserved. No part of this book may be reprinted or reproduced or utilized in any form or by any electronic, mechanical, or other means, now known or hereafter invented, including photocopying and recording, or in any information storage or retrieval sustem, without permission in writing from the publishers.

British Library Cataloguing in Publication Data
A catalogue record for this book is available from the British Library

Library of Congress Cataloging in Publication Data
Applied for

ISBN 978-0-415-08316-4 (hbk)
ISBN 978-0-415-75602-0 (pbk)

To Deb, Sebastian and Thomas

CONTENTS

ILLUSTRATIONS

PLATES

PLATE SECTIONS

[between pp. 6–7, 70–1, 120–1]

TABLES

ACKNOWLEDGEMENTS

This book means a great deal to me, but it would not have been possible without the help and generosity of numerous people and organisations. To all those who have contributed to making the book a reality, I wish to express my sincere thanks.

My work on facadism began in 1986 as part of a dissertation on the subject written during my Bachelor of Town Planning Degree course at Bristol Polytechnic. This book has grown out of that early work and I wish to acknowledge the contribution made by my tutor Tony McArthy with whom I had many fascinating discussions on the subject.

Many people in local planning authorities and private sector organisations have taken time to share their views with me and provide vital information, and my thanks go out to them.

I am especially grateful to my editor, Tristan Palmer, firstly for his faith in the project and secondly for his encouragement and assistance in providing vital leads which have enabled me to improve the book. I was particularly keen to include contextualising information from the USA in my work on facadism. Two people suggested by Tristan who deserve my thanks for their help in this respect are Greg Ashworth at the Faculty of Spatial Studies, University of Groningen, and Peter Larkham at the Faculty of the Built Environment, University of Central England, Birmingham.

The book has been greatly embellished by the work of Paul Guest, who produced some beautiful drawings for me, and he deserves a very special thank you.

I acknowledge with thanks permission granted by Methuen & Co. which has enabled me to use material from Roger Scruton's important book, *The Aesthetics of Architecture* (1979).

The maps in this book are based upon Ordnance Survey 1:2500 scale publications and are used with the permission of the Controller of Her Majesty's Stationery Office © Crown Copyright.

Finally, my special thanks to my family and friends for tolerating my preoccupation with this book over the last couple of years.

1

INTRODUCTION

My interest in the concept of facadism has its roots in the historic city of Bristol, where I worked as an urban planner during the 1970s and 1980s. The early part of this period brought drastic changes to the inner city, which saddened me. It was a time of mass destruction of areas of characteristic inner city townscape as a result of rampant commercialism, which in turn spawned a grotesque array of poor-quality modernist buildings, alien townscapes and spaces devoid of meaning, beauty or interest. Thankfully, the orgy of destruction came to a shuddering halt as the result of a mid-1970s property crash, which gave the historic built environment a reprieve and allowed time for people to take stock of what had happened to the centre of Bristol. Many felt a sense of outrage, and popular opinion provided a fertile land for the growth of new ideas about urban change. Suddenly, the urban conservation movement, which had been flexing its muscles during the 1960s, found a platform on which to base an effective and meaningful influence over the process of urban change.

During the late 1970s and throughout the 1980s, I became increasingly intrigued by the numerous cases around the inner city where old buildings had been completely felled – apart from the facades, which were held in place by elaborate scaffolding structures. These fragile-looking skins of brick or stone rose precariousy out of the ruins as monuments to the traditions of the past, survivors of comprehensively planned mass destruction. The bizarre and desolate scenes conjured up images of post-nuclear holocaust landscapes or displays from the world of avant-garde Punk art. In time, some of the facades became incorporated into new developments, with varying degrees of success. New infill terraced buildings behind a retained frontage were easily disguised, but a number of odd hybrids of old and new fabric appeared in more prominent situations. Other facades remained supported in place awaiting development, often for long periods, as the townscapes around them changed, creating equally strange scenes. Also at this time, the practice of building facsimile historic facades in front of new buildings was gathering pace in parts of the city.

Throughout Britain during the 1970s and 1980s, the trends in Bristol were being reflected in other historic cities. The approaches to redevelopment involving the preservation of historic facades or the creation of facsimiles in front of new buildings were greeted with controversy and labelled 'facadism', which has since been the subject of an emotive debate. Much disparaging comment about facadism has been generated from a wide range of interests involved in urban processes, including architects, town planners, historians and conservationists. For example, it has been claimed that facadism prevents new architectural styles from evolving and reduces buildings to 'mere elevations or self parodies'.[1] It has been condemned for causing the divorce between the interior and exterior of buildings and creating townscapes which are little more than stage sets. It is condemned by architectural purists as being immoral or distasteful. In the USA the practice of preserving facades as part of redevelopment schemes has been described in derogatory terms as façodomy by both the Modernist design community and conservation purists. However, justification has also been offered for facadism on the grounds that it is a valid method of urban conservation which enables the retention of familiar historic streetscapes or formal set pieces of urban design. At the same time, it also allows the provision of up-to-date accommodation with all its comforts and conveniences. Somewhat later than in Britain, the controversy about facadism has also surfaced in America, with similar arguments to those in the UK being rehearsed.

In 1986, when studying for my Bachelor of Town Planning degree, I was provided with an ideal opportunity to begin exploring the subject of facadism in depth. The material for this book builds on my initial research of that period, and work already carried out by others, although it must be said that I have unearthed relatively little written material.

Facadism is a concept which raises many significant implications for the built environment and its inhabitants. These implications provide the *raison d'être* for this book. It is important that concepts such as facadism are fully debated, explored, analysed and understood because they significantly embrace our lives and impact upon the quality of our day-to-day experiences. The association between the quality of human life and the built environment is evidenced throughout the history of civilisation by people's Utopian idealist strivings to provide ordered, dignified, beautiful and convenient settings for life's activities. The attractiveness and success of urban environments is influenced by many factors, such as the character, appearance and aesthetic qualities of particular townscapes, the accessibility of places to the community at large and standards of interior accommodation. Although Utopian urban areas may not be achievable, to be able to create the highest possible quality of environment the full implications of different approaches to urban development must be appreciated. I hope that this book contributes to such an appreciation.

It has been my intention here to broaden the perspective of the facadism debate both in terms of what is considered to constitute facadism and also in relation to the impact of facadism as an approach to urban remodelling. The debate has so far tended to focus rather too narrowly on the practices of preserving historic facades, or creating facsimiles, in front of contemporary buildings. There are other forms of development as well as facade preservation and replication which can, in essence, be regarded as manifestations of facadism, because they display similar characteristics or raise similar issues. Consequently, in this book I have considered, for instance, the time-honoured tradition of refacing buildings and the decorative exercises of postmodernism as part of the concept of facadism.

When we refer to facadism, what do we mean? Are we concerned with a conservation principle within which the preservation of street elevations is paramount or with an architectural philosophy wherein buildings are designed primarily with regard to their facades? Are we referring to a townscape concept which seeks to provide urban identity? Are we dealing with a marketing technique of heritage planning which aims to satisfy contemporary demands within the built environment?[2] Indeed, is facadism concerned with all of these matters to a degree? Certainly, facadism is an elusive concept which proves difficult to define, and one of my aims in this book is to explore ideas which bring the meaning of the term into sharper focus.

It is generally assumed that facadism is normally manifest in relation to historic buildings or townscapes. However, postmodern facadism, which the architect John Outram refers to as 'iconic' and 'pictorial' and as 'architecture recollected, not architecture itself',[3] is likely to appear on any street.

Architectural purists from any era would condemn facadism as a distasteful or dishonest approach to creating buildings. Others would argue that facadism is a time-honoured tradition practised by many of the world's great architects. I have examined facadism in the light of a range of architectural perspectives extending between these two extremes, under a threefold structure covering pre-modern, Modern and post-modern perspectives.

Facadism is also considered in relation to townscape philosophy and the three-dimensional visual impact of facadism on built urban form, and its inherent visual character and qualities are discussed.

Every development decision creates a balance sheet of social and financial costs and benefits involving many different interests. The effects of some forms of facadism can be seen as highly significant in terms of these costs and benefits. If we are to understand fully the consequences of concepts such as facadism as part of the complex process of urban change it is important to address such costs and benefits and their impact on various interests. Consequently, the book provides an overview of a number of relevant social and economic issues.

Facadism is also assessed in relation to the process of urban conservation. Here, the validity of facadism as an approach to remodelling historic urban environments and how true a concept it is to the spirit or essence of conservation are considered. Urban conservation is itself a somewhat elusive subject, but several basic themes are identified which help to define its essence. Having arrived at a conception of what urban conservation is, or should be, facadism is examined in the light of the main themes.

Finally, facadism is considered from a managerial perspective and explanations are provided of how and why facadism becomes a chosen development approach. The aim here is to contribute to a greater understanding of the nature of facadism as a course of action and its part in the achievement of urban remodelling objectives. The discussion in this chapter embraces the roles, attitudes and opinions of central government, local planning authorities, developers, architects, clients and amenity pressure groups within the complex negotiating processes that surround development decisions. There is also some analysis of the impact of particular outcomes on urban planning policy intentions.

It is, in my view, essential to an understanding of the concept of facadism to explore the subject more widely than simply assessing the end product of a highly complex decision-making process. Substantive issues such as facadism are not purely technical matters isolated from the complexities of the contemporary land development process and the political environment within which it operates. Courses of action within this process are subject to, and affected by, power struggles, negotiations and other interactions between a range of interests and the interplay of physical, social and economic factors. Therefore, in arriving at explanations of particular outcomes, I have examined the process that leads to the end product and have taken account of the values and aspirations of the various interests involved and the impact of relevant physical, social and economic factors.

Three different case studies provide the empirical analysis of particular outcomes. The explanations provided by these case studies clearly do not result in any universal truths about facadism. My aim is simply to highlight issues which may also be applied in other cases and thereby contribute to a greater understanding of the nature of facadism and its impact on the process of urban remodelling, particularly in the context of historic built environments.

The research methodology has involved a bottom-up approach to analysing the outcomes in the case studies displaying different characteristics wherein an action perspective, rather than a policy-centred or top-down appraisal, has been central to the analysis. Facadism may or may not emerge as a result of explicit or implicit policy strategies. Therefore, in my view, it is more appropriate to an understanding of the true nature of the concept of facadism to take particular outcomes, rather than policies, as the starting point of the analysis, and to examine the relevant influences that lead to those outcomes. The explanations arrived at provide an opportunity for reflection on the extent to which a course of action has related to policy objectives.

The analysis of the cases examined involved, where possible, open-ended discussions[4] with key actors in the decision-making processes and access to relevant documentation. Inevitably, personnel changes within organisations and an inability to trace individuals has meant that all matters could not be verified through discussion. Nevertheless, sufficient information has been compiled to enable adaquate explanations of the outcomes in each case to be reached.

A considerable amount of material in this book relates to Bristol, and I make no apologies for this. My detailed knowledge of the city and the relevant factors has enabled me to write with authority on many of the issues discussed. Bristol also possesses a number of good examples with which I have been able to illustrate certain points or arguments. However, there are also included numerous other cases from elsewhere in the UK to give the work a satisfactory balance. I have also included where possible contextualising material relating to facadism in Europe and the USA.

This is a book concerned with architectural and townscape philosophy, the economic and social effects of particular development decisions and the processes of urban conservation and heritage planning. It also attempts to unravel some of the complexities of the decision-making processes that lead to certain outcomes within the built environment. The book presents a collage of ideas relating to the concept of facadism which will hopefully enlighten what should be an ever-present debate about the built environment. If it achieves this aim and proves an enjoyable experience to the reader it will have been worthwhile.

Plate i Atrium Court, Glasgow.

Plate ii Interior, Atrium Court, Glasgow.

Plate iii Rear elevation of redeveloped building, Atrium Court, Glasgow.

Plate iv Restored facade, 1–6 Brunswick Square, Bristol.

Plate v Side elevation of redeveloped 1–6 Brunswick Square, Bristol.

Plate vi Brunel House, Bristol, 1992.

Plate vii(a) Detail of side elevation of Brunel House. See plate vii overleaf.

Plate vii Side elevation of Brunel House, Bristol.

2

WHAT IS FACADISM?

There is no universal agreement about what constitutes facadism and this represents a major difficulty in reaching an understanding of the subject. The most common use of the term is in relation to the practice of preserving historic facades or creating replicas, and the construction of essentially new buildings behind. There are also occasions when the facades of new buildings do not exactly replicate any other particular facades, but are designed with the intention of contextualising new development by evoking a typical style. However, I have encountered the argument that a distinction should be drawn between facadism – an approach to *new* buildings where the main elevation is designed as a component in a larger streetscape, and not necessarily as an expression of the building behind – and facade retention – a philosophy wherein an *existing* facade is re-used with a new building constructed behind.[1] I take the view that both approaches are, in essence, examples of facadism and, also, that if a broad perspective on the subject is taken, there are other forms of development which can be regarded as manifestations of facadism. For example, it can be argued that this is the case where the internal gutting and total redesign of freestanding buildings or buildings with more than one principal elevation leave only the outer shell of the building. Powerful townscape elements which become redundant, such as historic churches and warehouses, often have to be extensively remodelled inside to give them a viable new life.

While much of the debate surrounding facadism in the UK has centred on the conservation implications of the concept, a wider perspective introduces the argument that facadism may be manifest in buildings and townscapes of any age or type. Consider, for example, the facadism of early British postmodern architecture during the 1970s, in which contemporary buildings were dressed up in superficial Classical, Gothic or Art Nouveau motifs which related neither to the interiors of the buildings nor specifically to the surrounding townscape. In such cases the exterior decoration seemed to be simply an attempt to call up the past to solve the problem of making contemporary buildings more visually interesting and acceptable to the public.

The architect James Stirling is of the opinion that facadism is an element of 'compromise thinking in the UK'.[2] However, it is not simply a British phenomenon. There are many historic buildings and townscapes across Europe which have been remodelled through facadism. Helsinki Town Hall provides a striking example, and others exist in, for example, places as diverse as Bruges, Salerno and Paris. It is also a concept which has been adopted in many American towns and cities, with facade preservation being a particularly prevalent form. Numerous office redevelopment schemes have taken place behind the facades of older buildings in major cities, such as Washington DC, New York, Boston and Philadelphia. The practice has also occurred in smaller cities including Cambridge (Mass.), Kansas City, Salt Lake City and Portland. Old facades in Dallas, Brooklyn and Salt Lake City have provided frontages for new shopping stores. Parking lots and garages have been hidden by preserved facades in Charleston, Kansas City, Oregon, Portland and South Carolina. As in the UK, the concept of facadism has attracted controversy. In 1985, for example, the city of San Francisco, in response to lobbying from the Foundation of San Francisco's Architectural Heritage, adopted a plan which contained a designation to preserve 250 buildings in their entirety and 182 others in context. This was partly to counter proposals involving facadism, which was apparently considered an unacceptable approach to preserving the built environment.

It may appear, *prima facie*, that facadism is purely a postmodern concept, a symptom of the reaction against Modernist dogma and a compromise reflecting both the new spirit of urbanism and also the continuing need for urban areas to accommodate change. However, this is a restricted view of the subject and one that is challenged in this book because, although many aspects of facadism are postmodern in nature, the subject has much deeper roots in the past.

I have made my exploration of the concept of facadism through a diverse web of opinion, and arrived at a particular standpoint on the meaning of the term, which I hope will become evident as the chapter unfolds. The issues raised at this stage provide insights into the chapters that follow, where a more critical analysis of facadism is offered. In this opening chapter the task is simply to arrive at a conception of the various forms of urban remodelling that constitute a type of facadism and to recognise to which processes within broad urban systems facadism relates.

As a starting point, the approaches to development, or redevelopment, which are most often referred to as facadism, are identified below. These manifestations of facadism display several common characteristics and, by highlighting these characteristics, an important key to a greater understanding of facadism is provided. Also, having identified the characteristics, it is then possible to draw parallels with other types of development which can also be regarded, in essence, as facadism.

Almost every city and town in Britain, over the past fifteen years, has probably witnessed examples of intricate scaffolding structures – many of them fascinating engineering feats in their own right – supporting a delicate looking facade while the rest of the former building is demolished, never to return. Sometimes the scaffolding will have intruded menacingly into the street, blocking pavements and disrupting traffic. The technique is known as facade preservation or facade retention and is one of the most regularly occurring forms of facadism. Once the facade is supported *in situ* by scaffolding, a new building is constructed behind. Normally, the facade has no structural function in the new composition.

Considering the antipathy shown by many town planners towards facadism, there is a certain irony that in Bristol the home of the City Council Planning Department happens to be a modern office block, constructed behind the monumental facade of Brunel's former Royal Western Hotel, which dates from between 1837 and 1839 (see Plate Section, vi). A particularly interesting scheme in Glasgow City Centre carried out in 1989, at the corner of Gordon Street and Union Street, featured the retention of the facades of a cast-iron fronted building of Venetian inspiration called Ca'D'Oro, designed by John Honeyman and built in 1872 (see Plate Section, viii and ix). This project included the removal of a concrete mansard roof which had been added in the 1920s and the return of the facade to its original format. Another example in Bristol involved the facades of the whole of one side of a Georgian square being retained. In this case, at Brunswick Square, St Paul's, the original terraced houses, which had been converted since the last war to industrial premises, were replaced by a large office development behind the retained brick facade (see Plate Section, iv, v and xx).

A variation on this type of facadism occurs when a facade cannot, because of its poor structural condition, be preserved *in situ* while redevelopment takes place behind. Once the rest of the building has been demolished, the collapse of the facade is a very real possibility, despite elaborate scaffolding support. For instance, where bonding between bricks or stones is poor, or lintels over window openings have been weakened by decay, the facade may be structurally unstable. However, if it can be established before redevelopment begins that the facade to be retained is in poor condition, it can be carefully taken down and identically rebuilt. The importance of a thorough structural analysis of the condition of the facade, before any demolition of the rest of the building, is therefore paramount. Often, a considerable amount of the original material in the facade can be re-used, having where necessary been cleaned and restored. A visit to the North Devon market town of Barnstaple reveals such an example. At the corner of the town's High Street and Joy Street stands a retail development nearing completion where a facade to Joy Street was taken down and rebuilt, mostly using the original bricks and dressed stone (see Plate 1). Some replacement of features of detail was necessary, but these were fashioned to match the originals.

Where excessive damage has occurred to original materials either through environmental damage or the demolition process, making them unusable, and a decision has been made to retain a particular street frontage as part of a redevelopment scheme, the construction of replica facades in appropriate new materials provides a solution. Another manifestation of facadism occurs when a new development is fronted with facades which are not exact replicas of the facades of former buildings on the site, but simply evoke a particular style. In Portland Square, Bristol, a number of the facades are Georgian replicas which were constructed in new Bath Stone in the late 1970s and early 1980s. Just around the corner, in Pritchard Street, is a rank of buildings erected in

Plate 1 Joy Street, Barnstaple, North Devon. A rebuilt facade stands in front of a new retail development.

the late 1980s, with facades which evoke the Georgian style of original buildings nearby. Across the City Centre is Queen Square, built between 1699 and 1727, the first and largest eighteenth-century square developed in Bristol. In the south-west corner of the square, two late-Victorian warehouses were demolished in the late 1970s and replaced by new buildings. Immediately to the east of this development four houses – probably originating from the early 1700s, two of which were refronted in the nineteenth century – have since been replaced as a result of a failed attempt to preserve the original facade, which was insufficiently strong to survive such radical treatment. The facades of these new buildings fronting the square attempt to evoke a group of early eighteenth-century terraced houses. Behind the facades are not terraced houses but open-plan offices. There are no party walls behind the quoins. The rear elevation of the building is of an entirely different style and one which tries to reinterpret the Bristol warehouse vernacular. Also the pitch of the roofs facing Queen Square is radically different to that of authentic houses in the square, built in the early 1800s.

All of the forms of facadism mentioned above display a number of common characteristics, one of the most controversial being that it leads to buildings where there is little or no relationship between the facade and the rest of the building in terms of style, proportion and structure. There occurs, therefore, a divorce between interior and exterior. Reflecting again on one of the Bristol examples mentioned above, this aspect is well illustrated: in Queen Square we have the terraced buildings erected in the late 1970s with the facade onto Queen Square evoking early eighteenth-century houses, the rear facade onto The Grove reinterpreting vernacular warehousing and, inside, open-plan office space stretching along the entire length of the buildings. The party quoins to the Queen Square elevation suggest vertical subdivision inside the building which, of course, does not exist in reality, the illusion being shattered on close inspection by the strip lighting and lack of interior dividing walls. Further, the mock eighteenth-century facades imitate the load-bearing walls of the original buildings of that period. However, they are in reality only decorative skins of four-inch brickwork.

The late 1980s saw a prominent element of Classical revivalism emerging in Britain, one of the main protagonists being Quinlan Terry, who has constantly striven, even during the days of rampant Modernism and then High-Tech, to continue the English Georgian townscape tradition. Parts of his Richmond riverside development in Surrey are a pot-pourri of eighteenth-century-style facadism. Fronting the Thames, the scheme comprises a mix of brand new, restored and remodelled Georgian houses, civic and commercial buildings. It is an important townscape exercise in that it demonstrates a contextual approach to urban design within the constraints of having to accommodate conventional contemporary interiors within the new buildings. Therefore, for the most part, in Terry's new buildings at Richmond, Classicism stops at the highly mannered facades, behind which are the common features of present-day office accommodation such as suspended ceilings which drop below window heights, air conditioning plant and strip fluorescent lighting.

With the characteristic divorce between exterior and interior in mind, it is interesting to focus for a moment on a French example, which demonstrates that facadism is not simply a British phenomenon. The Taller de Arquitectura is one of the most unusual and successful architectural practices in the world. It was founded by Ricardo Bofill, a Spaniard born in Barcelona in 1939. Bofill made his reputation by designing and building housing schemes in the 1970s in the Paris suburbs at Marne-la-Vallée to the east of the city and St Quentin-en-Yvelines to the west. Externally, the buildings appear as colossal classical palaces set in massive imperial Roman squares, crescents, circuses and piazzas. However, behind the facades, classicism evaporates and the apartments are similar to most others found around western Europe that were built at the same time.

A further closely related characteristic of facadism concerns its impact on the grain

Plate 2 The historic High Street area of Barnstaple, North Devon provides an example of close-knit townscape grain.

or cellular structure of townscape. The grain derives from, for example, the plan form of buildings and existence of party walls which divide properties, the width of building frontages, whether plot sizes and shapes are regular or otherwise, and also the density of development. Such factors combine to form the cellular structure of townscape. The Shambles in York, the centre of St Ives in Cornwall and the High Street area of the North Devon market town of Barnstaple (see Plate 2) are examples of parts of settlements possessing a very tight, uneven grain made up of closely knit building fabric, irregularly shaped property plots, and diverse patterns in plan forms, styles of buildings, textures and colours. In comparison, the Georgian masterpieces of Royal Crescent, the Circus and Queen Square in Bath (see Plate 3), or parts of Edinburgh's First New Town display a much more open grain with an ordered pattern of development at a lower density, less diverse plan forms, and relatively fewer variations in style, colours and textures. Facadism can drastically change the grain of townscape, and thereby its character. If, for instance, behind a preserved facade, where there once stood a terrace developed on irregularly shaped plots, all the party walls of the former buildings are removed and the plan form changed to achieve a new building of open-plan design inside, the grain of that part of the townscape will become wider and take on a different character.

A further characteristic of facadism is that it results in the loss of historical information contained in the fabric behind facades. Two cases in Bristol of 'what might have been' had the developers' original proposals been implemented illustrate this point. First, in Redcliffe Street, a developer was proposing to keep a facade fronting Redcliffe Street and redevelop behind this, where it had been found that the fabric contained part of the old city wall. The historical significance of this is obvious and the wall has since been incorporated into the redevelopment. In the second case, in Victoria Street, Bristol, in the mid-1980s a developer produced a scheme which involved the removal of medieval fabric behind Victorian facades. The buildings were found to contain a great deal of valuable information about the old street pattern and house plots before the Victorian road-building programme in the area. Lengthy negotiations have ensued between the developers' architects and the city planners and to date all the fabric still remains, albeit in a derelict condition.

Less obvious perhaps than other characteristics is the fact that facadism simultaneously contributes to the protection, at least to a certain extent, of the physical character of townscape, and also to significant changes in the social and economic structure of urban areas. On the one hand, familiar street scenes and landmarks may be retained or, in the case of replica facades and those which evoke previous styles, what might be regarded as externally contextual development may be created. However, at the same time, an established type of accommodation is removed and replaced with another and, as a consequence, there is likely to be a displacement of traditional occupier types and changes in property sale and rental values. This is not to say that facadism is either a cause or an effect of the process of urban change. However, it is inextricably linked with that process. Facadism can provide a mechanism by which

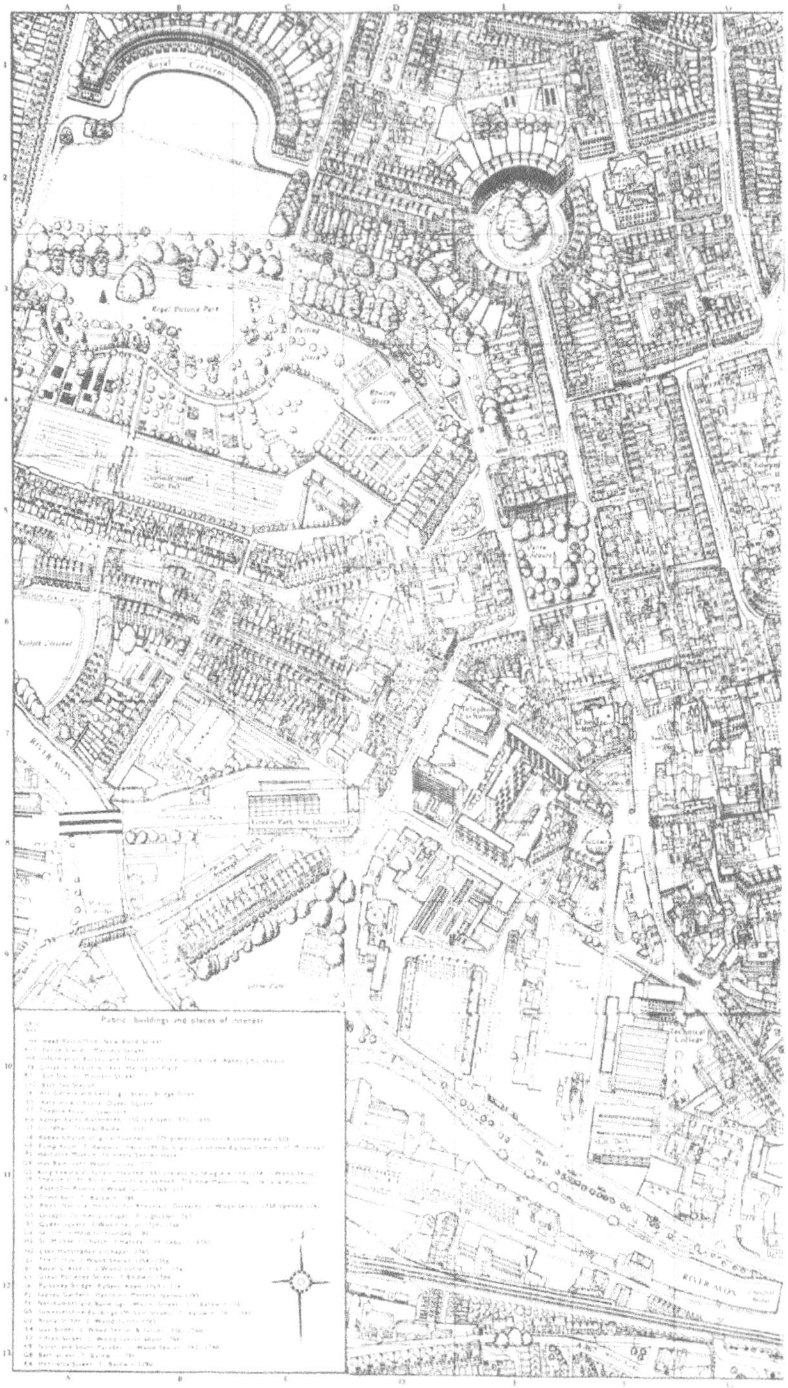

Plate 3 A pictorial representation of the city of Bath: wide or loose townscape grain.

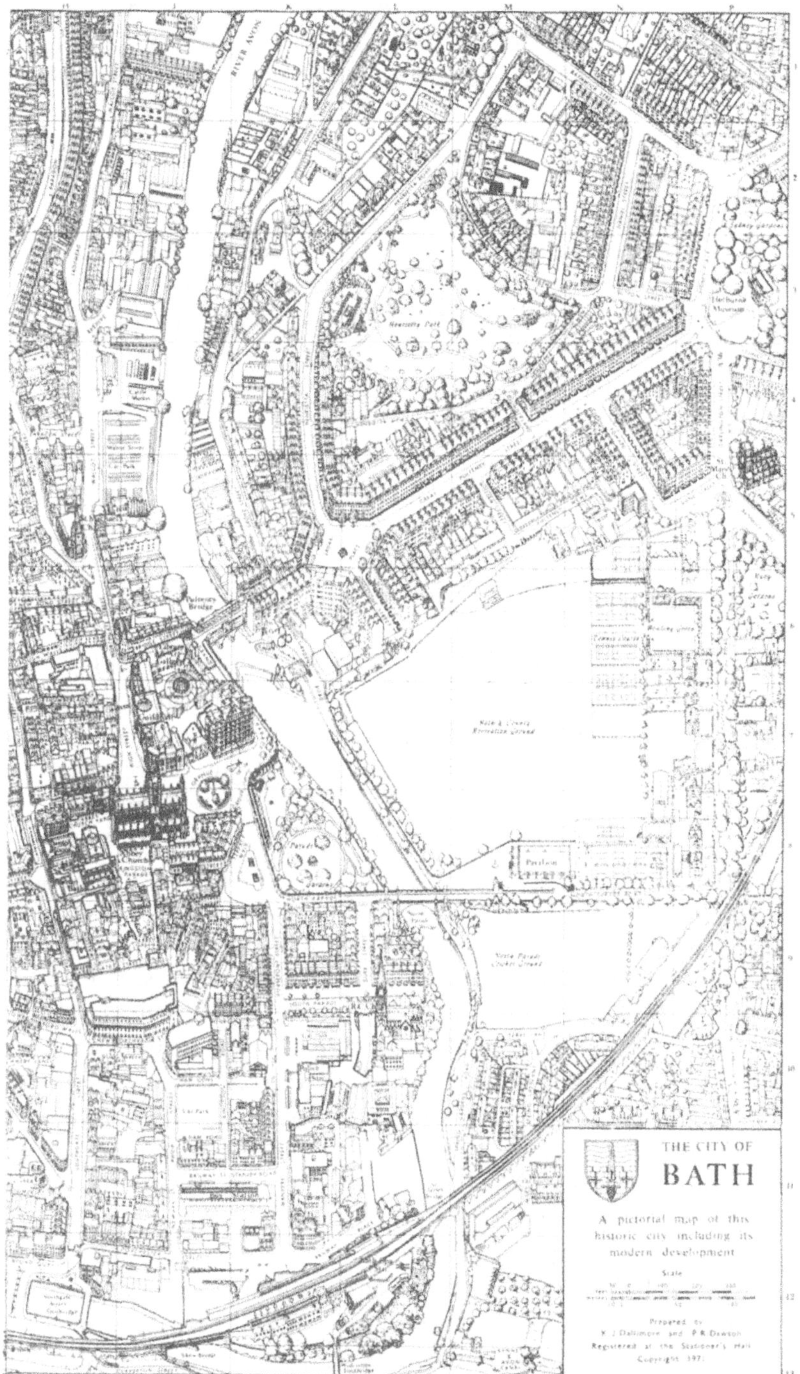
THE CITY OF
BATH
A pictorial map of this historic city including its modern development
Scale
Prepared by
K J Dallimore and P R Dewson
Registered at the Stationer's Hall

developers are able to meet, at least in part, the objectives of urban conservation interests by achieving the retention of streetscapes. At the same time it also allows developers to replace accommodation which may not meet their marketing requirements, perhaps because of its size, style or image, with buildings more adaptable and appropriate to their needs.

Thinking hypothetically for a moment to illustrate this point, a likely scenario might be as follows. A developer owns a dilapidated terrace of Georgian houses, in a designated conservation area, near a town centre. In the area, the process of displacement of the resident population by commercial interests has begun and some former houses have already been converted to office use. The developer's feasibility studies indicate that, with the area being close to the town centre, there is considerable demand from larger commercial interests for up-to-date open-plan office space, but the refurbishment of existing buildings for either residential use or small subdivided suites of offices would make the investment far less secure. A scheme of facade retention, with the rest of the existing buildings behind being cleared and replaced with purpose-designed accommodation would be likely to achieve the developer's prime objective of maximising profit and may also go some way to appeasing local conservation interests.

The developer's proposed facade retention scheme could result in a number of social and economic changes occurring in the area. The process of the locality becoming part of the commercial sector of the town, rather than remaining a residential area, would be further cemented. Also, the removal of the traditional fabric behind the facade would make this transition seem more permanent than if the developer were compelled to convert the buildings in their existing form. The latter approach might convey the thought that the buildings could one day return to their former use, whereas if most of the buildings have been replaced such a thought seems irrelevant.

Depending on market conditions at the time of development, the constraints of having to convert buildings in their existing interior form could result in less viable office space being created, because the type of accommodation provided by the conversion scheme might not reflect what is in demand. The replacement of smaller-scale accommodation by modern office space would open the way for larger commercial interests to infiltrate the area, with a likely consequence of significantly enhanced rental and property values. The local retail economy could conceivably receive a boost from greater numbers of office workers being brought to the area and thereby increasing spending power. Also, the provision of open-plan office space would probably increase the development plot ratio and number of people occupying the site with a consequent increase of traffic, and greater demand for car parking in the locality with the accompanying social costs of congestion and pollution.

Turning to an actual example of Blythswood Hill in Glasgow, within an area of late Georgian terraced housing lying on the edge of the early nineteenth-century city, and an area which constituted Glasgow's second New Town, many of the original domestic buildings have over the years been occupied by more intensive commercial uses. More than a dozen facade retention projects had been carried out by the end of 1990, with new office space being provided behind the facades. One conclusion that could be drawn in this case is that facadism has facilitated the change from residential to commercial area and the consequent impact on the social and economic structure of that part of Glasgow, by offering a compromise development solution which has satisfied some conservation objectives and, at the same time, allowed the removal of many of the vestiges of past social and economic orders.

While the above characteristics relate to the most commonly referred to manifestations of facadism, certain parallels can be drawn with other approaches to development, and it can be argued that they are, in essence, forms of facadism. At least they need to be spoken of in the same breath as facadism! For instance, what else, other than a form of facadism, is the complete internal gutting and redesign of freestanding buildings such as redundant churches or former warehouses?

The Bush Warehouse in Bristol City Docks actually has three facades facing public places, but offers an excellent example. Here, the three outer walls and the party wall were kept and the roof and interior redesigned to create an interesting Arts complex (see Plate Section, xiv). The characteristic divorce between interior and exterior found in facadist schemes in terms of form and function is evident. Even though the basic form of the warehouse at street level remains the same when the building is seen at close quarters, the changed roof design of the converted building has had a considerable impact on the character and appearance of the townscape when seen from a distance. Also, the radical degree of internal change to the building, which has allowed the creation of the Arts complex, has furthered the social and economic change that has taken place in Bristol's City docks. A less drastic internal remodelling of the original building may possibly have constrained the creation of a facility which draws large numbers of people to the waterfront, and one which has been a major element in cementing the growth of cultural and leisure uses in the docks. If, for instance, a less radical solution had occurred, and the internal layout of the original warehouse had been reflected in an office conversion scheme, which given past demand for office space in the inner city may well have been a likely scenario, the social and economic transformation of the dock area would have been quite different. Like the earlier examples of facade preservation, creation of replicas in front of new buildings and facades evoking traditional styles, internal gutting and remodelling provides the flexibility for much more extensive social and economic change.

Church buildings do not lend themselves to sympathetic conversion to other uses. The basic cruciform of many traditional redundant Anglican churches is a major constraint in structural and aesthetic terms. Nonconformist churches of simpler form tend to lend themselves more easily to radical internal remodelling. Whatever the proposals to convert churches to other uses, the resulting divorce between interior and exterior is highly significant. Church buildings perform a very significant role as landmarks and symbols within townscapes and convey many deeply meaningful messages. In the event of churches being declared redundant and converted to alternative uses, providing their exterior form remains intact, their landmark role can continue. The change may, however, have some impact in terms of symbolism. The exterior form and appearance of a church is a strong outward symbol of its interior form and function. It is because churches provide powerful images of the cycle of human life and beyond that the link between the physical form of the exterior of the buildings and their interiors, and also the activities that take place within, is so strong. The symbolism and associations attached to churches play an important role in assisting people to identify with and understand the built environment. Even though redundant churches may be given new life, a radical internal redesign of a church building to accommodate a different activity brings about what may be for many people a traumatic divorce between interior and exterior, and one which confuses the messages and cues projected by the building.

In conceptual terms, there is a comparison to be drawn between the internally remodelled church and a facade preservation scheme which, from the street reads as a terrace of houses, but is in fact open-plan offices inside. The messages given by the facade to people on the street do not correspond with the use of the building and its internal form. Also, the internal conversion of a disused church, as with other forms of facadism, provides a clear indicator of the process of social and economic change in an area. It is accepted that a simple change of use of the building would also provide a signal of this process, but the actual physical remodelling of the interior conveys the message that the change is much deeper seated and permanent. Further, one of the aspects of facadism highlighted earlier in the chapter was the possible loss of important information if all the fabric behind a facade is removed. The same may equally apply if a church is gutted and remodelled inside.

The time-honoured tradition, used by a number of great architects and designers, including Palladio and Wren, of replacing or hiding the facade of an existing building behind another should, in my view, be included in a consideration of the subject of facadism. At Hampton Court, for example, Wren inserted classical fronts onto gutted medieval buildings and Palladio reclothed the great fifteenth-century Palazzo della Ragione (Basilica) in the centre of Vicenza by forming a classical colonnade in front of the medieval market hall. More mundane examples of refacing buildings are to be found in most towns and cities in Britain dating from many different eras; in particular, the practice was popular in Georgian times. Earlier in this chapter two examples in Bristol were highlighted from different periods. At Queen Square, two early eighteenth-century houses were refronted in the following century and, during the Victorian era, at 10–22 Victoria Street new facades were constructed, including examples of the Bristol Byzantine style, in front of medieval buildings (see Plate 4).

The practice of giving buildings new facades has not been confined to the history books. G. J. Ashworth refers to 'the ultimate exercise in "facadism"' in his book *Heritage Planning*, in relation to a bizarre example from the small town of Yakima, Eastern Washington, in the United States, where a set of facades, representing the popular fantasy of a western town, has been built in front of a converted warehouse used as a shopping mall.[3]

Plate 4 Victorian detailing of the Bristol Byzantine style on a facade at Victoria Street, Bristol, behind which and others in the terrace are the remains of rare medieval buildings.

The practice of refronting buildings reflects some of the characteristics of other forms of facadism. The conceptual divorce of the exterior and interior of the building is as likely to occur in cases of refronting as with facade preservation or replication. Also, cues and messages to people on the street may be inaccurate. What has the wild west facade at Yakima to do with a twentieth-century shopping mall other than the fact that the jostling shoppers may feel that they are in fact involved in a saloon brawl? In the case of 10–22 Victoria Street, Bristol, the nineteenth-century facades give no hint of the form or the age of the medieval fabric behind. While refronting in some respects appears to be a directly opposite approach to facade preservation or replication, in the sense that it is seeking to change streetscapes rather than retain the status quo, it is conceivable that refronting could be used to contextualise a building better with its surroundings. The objective with both forms of development would therefore be the same, i.e. the enhancement of the character of townscape.

In architecture, postmodernism has resulted in a historical revival, with many new buildings looking to an array of past styles for inspiration. Early attempts at Classical revival included facades with stuck-on pediments and painted columns. The Terry Farrell Partnership produced an office block in 1986 in Queen Street, London, where the language of Art Deco is powerfully reflected in the elevations in the style of the great New York skyscrapers. The detailing of the base of the facade also shows a Viennese link with the Secessionist architect, Otto Wagner (1841–1918). It is clear, however, from the facade's detailing that the exterior panels are not solid, but have been clipped on, and in this respect the building is unpretentious. Inside, the building is straightforward in terms of its plan and accommodation. In 1987, Chapman Taylor and Partners proposed a shopping centre constructed in the form of a colossal Palladian villa. The place of facadism within the postmodern era is dealt with in the next chapter, but clear parallels can be drawn at this stage between techniques such as facade preservation, the building of facades in replica or those which mimic or evoke a past style, and some forms of postmodern architecture. First, there is a reliance on the past to solve the problem of the facade. Second, there is an absence of conceptual unity in terms of style and form between the exterior and interior.

Having explored the different forms of urban remodelling, which may, in essence, be regarded as examples of facadism, and considered their characteristics, it is now possible to arrive at an idea of the types of processes with which facadism is connected.

First, the conservation dimension is considered. This is the process with which facadism is most commonly associated and the various manifestations of facadism raise many implications in terms of the achievement of urban conservation objectives (see Chapter 6, 'Facadism and Urban Conservation'). Since the early 1970s, one of the major preoccupations of the urban conservation process in the UK has been to preserve and enhance the character and appearance of elements of the built environment, which are perceived to be valuable in architectural and historic terms, and to give them continued life within a process of urban change. These valuable elements include whole areas of townscape, small groups of buildings, individual buildings and parts of buildings. The facades of buildings may have intrinsic aesthetic qualities or historic associations which merit their retention irrespective of the rest of the fabric to which they are attached. Also, it may be felt that, in the event of redevelopment, the character of certain historic areas may be best preserved or enhanced by the new buildings having facsimile facades, or facades evoking a particular traditional style. The question of what happens behind these facades raises a range of significant conservation issues. So does the matter of internally remodelling and re-using obsolete historic buildings. Such buildings may be vital to the character and appearance of areas, and their internal redesign and conversion may be the only means of ensuring their survival. Further, the use of particular stylistic features on the facades of postmodern buildings is one approach to trying to contextualise new development within historic townscapes.

Facadism also embraces architectural philosophy. Take, for example, the facadism of postmodern architecture which the architect John Outram claims 'is pictorial', but is 'architecture recollected not architecture itself'.[4] Also consider the art of architectural compromise involved in the preservation or replication of facades as part of urban remodelling schemes. In some cases, successfully linking a new building to a historic facade presents an immense architectural challenge. This is particularly so when the building is exposed to public view rather than set within a terrace. Further, facadism may be seen as a time-honoured architectural tradition in that, in many periods of history, buildings have been refaced or reclothed in a new style.

'Townscape' is another philosophy to which facadism is inextricably linked. The facades of buildings make an important contribution to urban views, vistas, streetscenes and other townscape features. They may also be vital elements in creating not only a sense of place but also the physical and spatial qualities that contribute to the identity of some urban areas. Decisions about whether or not to preserve, rebuild, replicate or replace facades and the nature of development behind facades are of fundamental importance in terms of townscape quality. Facadism is generally associated with historic townscape, but it is also relevant to consider it in the context of the whole spectrum of urban form. The facadism of postmodern architecture, for instance, may relate equally to completely contemporary townscapes as to historic areas.

So far, we have seen that facadism emerges within an overlap between the urban conservation process and architectural and townscape philosophies. Finally, facadism must also be considered in the context of another process which G. J. Ashworth calls 'Heritage planning'.[5] In essence, heritage planning is a synthesis of three fields of thought:

> These are, first, the preservation and presentation of the surviving buildings, relics, memories and place-associations from the past; secondly, the contemporary industry that uses these historic resources to satisfy modern demands . . .; and thirdly planning intervention to rehabilitate and revitalise local economies and communities.
>
> (Ashworth 1991)

A heritage planning perspective sees all conservation as a 'market driven set of choices about what will satisfy contemporary demands'.[6] It is, therefore, a demand- rather than supply-oriented process. The practice of retaining existing facades or building replicas can be seen as one of satisfying current demands for both the preservation of familiar traditional settings for life's activities and also the provision of the required standards of modern accommodation. Further, facadism is also a mechanism for using a combination of heritage – in the form of historic facades or old building shells – and contemporary development to revitalise local economies and communities.

Inevitably, this chapter has provided only the starting point in answering the question 'What is facadism?' and the following chapters will, I hope, provide the reader with deeper insights into the meaning of the concept.

3

ARCHITECTURAL PERSPECTIVES

Ever since its emergence within architectural and town planning vocabulary during the 1970s, the word facadism has stimulated controversial comment in a complex debate about its validity and morality as an approach to development. This debate highlights an enormous gulf between differing opinions about the nature of buildings and the architectural significance attached to 'the facade' as an element of building design. From some perspectives, facadism presents an unwelcome challenge to the conceptual foundations on which the principles of good architecture are based. Architectural purists from a number of different 'schools of thought' would regard facadism as dishonest and distasteful, claiming that it leads to 'sham' buildings. From another perspective, however, it may be argued that facadism is justified by historical precedent. Also, facadism may be seen as part of a continuing trend, certainly evident in Britain since the mid-1970s, towards a re-examination of history in architecture and the popularity of 'contextualism', and as a particular current fashion in building design. An important point not to overlook is that a number of the arguments raised involve not only matters of architectural style and design, but embrace a range of conceptual issues based on physical, social, political, economic, philosophical, religious and historical considerations.

Some of the views expressed in the debate have been taken from emotive or subjective stances. Others are more objectively based in that they reflect an analysis of facadism in the light of widely acknowledged, if not universally accepted, architectural principles and ideas. In drawing this distinction, no attempt is being made to devalue subjective opinion, which is often expressed both passionately and knowledgeably, but simply to highlight the fact that some views are based more on rational analysis and others more on personal taste.

The issues raised in the facadism debate can be linked to a number of different eras and 'schools of thought' throughout architectural history from classical antiquity to the present postmodern age. Primarily to provide some sort of coherent structure to the chapter, I have analysed the concept of facadism from three perspectives: pre-modern, Modern and postmodern. The ideas discussed cannot be compartmentalised completely, however, given that throughout architectural history certain ideologies and concepts have, from time to time, been revived or re-examined and reworked. For example, the Greek notions on proportion displayed in The Golden Section find an echo in Le Corbusier's Modulor. Also, in relation to Modernist perspectives on facadism, while accepting that some of the ideology has become discredited and a new era has been born, there is still a broad Modern church, carrying forward the original threads of thought, within which there has been logical and consistent development of ideas embracing more pragmatic and experimental forms of Modern design, as exemplified by Norman Foster and Richard Rogers.

PRE-MODERN PERSPECTIVES

One of the common criticisms raised by opponents is that facadism causes a lamentable divorce between the interior and exterior of buildings. Such purist analysis of building form expresses a somewhat ascetic viewpoint that the exterior of a building should always be a natural expression of its interior. This is indeed an ancient concept with discernible origins in classical antiquity and the greatly respected Vitruvian concept of harmonic proportion, later reflected in the High Renaissance architecture of Alberti and Bramante. To Vitruvius (first century BC) the creation of beautiful architecture required the proportions of all parts of a building to be rationally integrated, with every part having a fixed size and shape; to add or take away anything would destroy the harmony of the whole composition. As a concept, facadism, whether in the form of refronting buildings or of replacing accommodation behind retained and facsimile facades, emphasises the potential for treating the exterior and interior of buildings as separate wholes, and is at odds with the classical ideal. It would be possible, for instance, to refront a building, or create a new interior behind a retained facade, in a form and style which expresses the idea of harmonic proportion. However, an important aspect of facadism is that it provides an opportunity to give a building a completely new face – perhaps to follow a current fashion – or to radically redesign an interior which may have become obsolete.

Within the facadism debate, it should be recognised that many buildings, now regarded as architectural treasures, are not the result of architects rigidly applying established rules or dogma. It has been said that 'all the great architects are guilty of facadism, pastiche and replica',[1] if, indeed, there is any reason for guilt. Even SAVE, an influential conservation amenity group, who fiercely reject the validity of facadism nowadays as an approach to building design, recognise that it 'has always in a sense, been an element in architecture'.[2] The conceptual separation of exterior and interior may be perceived in different forms, and has been evident over time both in the adaptation of buildings – which has often involved the fusion of different styles to form new compositions – and in the construction of completely new buildings.

Consider, for a moment, a piece of work by one of the world's most revered architects, Andrea Palladio, who in 1549 reclothed the Palazzo della Ragione (Basilica) in Vicenza, by erecting a classical facade in front of an earlier medieval market hall.

Here, he used his unquestioned ingenuity in solving the architectural problem of the whole composition, but the outer vesture, the facade, was also given an intrinsic value of its own. In his work generally, Palladio rethought the Renaissance traditions of ancient Rome and, in the Mannerist fashion of Michelangelo, intentionally broke the ancient classical rules in creating an original effect. The reclothed Vicenza Basilica provides an example of the conceptual separation of facade and interior, and, it may be argued, is in effect a form of facadism. Normally, nowadays, facadism involves the retention of a historic facade of intrinsic value, or the creation of a new facade in a traditional style, as the front elevation of a modern building, rather than the refronting of buildings. However, it still involves the fusion of old and new fabric to form a new composition, and a recognition of the intrinsic value of the built facade, as in the case of Palladio's Basilica. Given that one of the greatest of all architects was not averse to breaking established architectural rules in creating original effects, there is every justification for present-day architects pioneering different ways of achieving an objective. In the case of facadism, there appears to be no reason why traditional facades cannot be creatively incorporated into 'state of the art' developments. Indeed, the task of knitting old and new should be a stimulating challenge. A case where the challenge has been taken up, and successfully met, is at Atrium Court at West Campbell Street, in the centre of Glasgow. From three of the four streets which surround the building one is faced with the heroic, early twentieth-century stone facade. Passing through the facade into the new airy atrium the effect is that of an uplifting surprise, or a pleasant shock which quickens the pulse, such is the dramatic contrast between the streetscape and interior (see Plate Section, ii).

Another concept relevant to the facadism debate is that of the urban facade as architectural scenery. This was certainly evident in Baroque architecture and town planning in Europe, spanning the sixteenth to eighteenth centuries. The period saw a close relationship between stage design and town planning, and Baroque facades were designed to impress during promenades and ceremonials and often conformed to laid-down design codes. In Paris, the Place Vendôme was built during the early eighteenth century, and here the facades were constructed at the same time to a uniform design, the rest of the building fabric being added later. As Zucker comments: 'This procedure proves the absolute priority of artistic purpose over any functional demands'.[3] In Britain this idea can also be found in the work of Inigo Jones in the seventeenth century, in much Georgian set-piece development and in particular the work of the Wood family in Bath during the eighteenth century. Also, in Berlin, where the construction of streets having uniform facades and focusing on *ronds-points* – in the French style – was adopted, and in Paris, later in the nineteenth century, where Baron Haussmann proceeded to lengthen many of the facades in a late classicist style. Some would say that he did so 'to the point of vulgarity and boredom',[4] but others may feel inspired by the vastness and unity of the streetscape.

Turning to the terrace architecture of Inigo Jones, the Wood family in Bath, and John Nash, we find further clear illustrations of architectural scenery, where the urban facade has been designed with respect to the townscape rather than the accommodation behind. It also emphasises the point that the urban facade, in design terms, may relate, in some cases, not to a single building but to a whole street frontage or other planned format. Inigo Jones experimented in this country in the seventeenth century with the concept of giving a uniform Palladian facade to a street of houses or bounding a square. The practice lapsed after Jones's time but was revitalised in Bath in the early eighteenth century by John Wood. He envisaged town houses not as individual units, but as part of an overall planned architectural layout for the city. Between 1728 and 1775 John Wood and his son (also named John) completed the development of a magnificent townscape composition, including Queen Square, the Circus and Royal Crescent. The uniform stone facade of Royal Crescent features 114 Ionic columns and stands in front of 30 town houses. The statement made by the facade is not inward looking – that is, it is not conveying the idea that there are 30 individual houses behind – but is a bold outward expression of an elegant set-piece design, which many would say constitutes one of the most impressive curving facades in Britain.

The validity of facadism in relation to set-piece design is one of the fundamental aspects of the debate on the subject. Standing within a formal Georgian Square you are left with the feeling that to drastically alter any one of the facades would be akin to ruining an elegantly proportioned and tastefully decorated room in a house. In the 1980s, at Brunswick Square in Bristol, a facade of an otherwise totally derelict terrace of buildings, forming one whole side of a formal Georgian Square, was preserved (see Plate Section, iv). The accommodation behind the facade is part of a vast modern open-plan office complex, and in reality the old and new fabric sit uneasily together. The terrace of former Georgian houses at 1–6 Brunswick Square in Bristol still read as such from the street, but, in fact, behind the retained facade with its party quoins, there are no party walls and the rebuilt chimneys are false, these being supported by steel beams in the roof space. This is a classic example of one of many cases of facadism which can be condemned by opponents as dishonest and distasteful. There is deception in that the party quoins do not reflect the interior space and the chimneys are purely decorative features on the roof. Yet it must also be recognised that the facade in its own right has considerable architectural merit and its retention has preserved a most pleasing piece of urban design. It is also true that the retention of such an example of pre-modern design currently receives wide popular support. However, further consideration of the townscape aspects of facadism must wait until the following chapter.

When buildings such as Palladio's Vicenza Basilica, those of Jones, the Wood brothers and Nash, the grand Baroque facades and many more examples, too numerous to mention in this book, are considered, there can be little doubt that distinctive features of fine buildings of the past have obtained a perennial value, and one such feature is 'the division between inside and outside, and the consequent vesting of architectural significance in a built facade'.[5] This being accepted, there is then considerable weight in the argument that various forms of facadism are justifiable in some instances, such as, for example, where an attempt is being made to complete unfinished architectural scenery, or extend it, or to ensure the retention of a set-piece composition; also in cases where modernisation of a building is necessary and the facade is the only element of architectural value. This theme will be taken up again in considering facadism from a Modernist perspective.

Attitudes towards the facades of buildings since the mid-1970s in Britain show certain similarities to the desires in past eras to create architectural scenery. A strong motivation for either retaining historic facades, or building new facades which evoke or replicate a particular past style, sometimes emanates from architects whose clients wish, through the facade, to project a particular image. For example, an organisation may feel that an elegant Georgian facade suits its image more than a Modern or postmodern exterior. Perhaps a heroic classical-style facade is more in keeping with another company's self-image. Not uncommonly, the outcome is a form of facadism, either through the adaptation of buildings, by providing new accommodation behind a preserved facade or the construction of a replica facade, or by a facade evoking a required style, in front of a modern structure. The aim is to make an outward gesture or statement, designed to impress, inspire or influence those who see the building from the street.

As mentioned earlier, the criticisms of facadism concerning the divorce caused between the interior and exterior of buildings display functionalist thinking. Such views have links both with the Modern movement and the subtle functionalism displayed by A. W. N. Pugin (1812–52) and his followers in the nineteenth century. A consideration of facadism in the light of the functionalism encountered during the Modern era follows later, but at this point it is important to reflect on Pugin's views and work because they represent a most important watershed in terms of architectural history, and particularly on his significant contribution to a much greater and more serious understanding of the functions of individual parts of buildings and their interrelationship with structure, style and decoration. Nowadays, some influential amenity groups who campaign for conservation of the built environment, such as SAVE, claim to follow 'the Pugin line'[6] and argue that facadism encourages the tendency to regard buildings as mere frontages, and creates sham buildings. It is true that one of the characteristics of many manifestations of facadism is the lack of any relationship between the facade and the rest of the building in terms of structure, style and proportion. However, before dealing with this aspect of facadism, it would be useful to provide some contextual background to Pugin and his contemporaries.

Pugin captured the public attention in 1836 in a famous book entitled *Contrasts*[7] wherein he fiercely condemned the contemporary architecture in comparison with buildings of the Middle Ages, which he regarded as the 'greatest era in human history, a time when men were inspired to live and work for the glory of God and when buildings such as the great cathedrals were erected in this spirit'.[8] His conversion to Catholicism in 1834 clearly influenced his thinking on building design and provided considerable motivation in his work. This is evident in *Contrasts* and a later book *The True Principles of Pointed or Christian Architecture* (1841). He upheld, as an icon for his time, the late thirteenth and early fourteenth centuries, which to Pugin were the greatest days for architecture and, indeed, for Catholicism. Clearly the characteristic vertical emphasis of Gothic architecture, with its lofty spires, flying buttresses, pointed arches and ribbed vaulting, and an apparent reaching upwards to heaven, reflects Pugin's religious stance, but another important factor was the structural honesty of the Gothic style. The resultant structure is delicate yet resilient, wherein each part is dependent upon and related to the others.

Pugin's zealous pursuit of establishing Gothic Revival as a serious style succeeded. At the same time he was heaping ridicule on Renaissance architecture, which led, in his view, to absurd examples such as the dressing up of a house as a classical-style temple. Pugin viewed such architecture as being part of 'the Pagan Revival'. He equally condemned sham Gothic architecture such as that shown in the drawing in Plate 5 which is based on an illustration from *The True Principles of Pointed or Christian Architecture.* Here, a grand facade stands in front of a simple brick building. To Pugin and the medievalists, reference to function was 'necessary as a standard of taste, a means of distinguishing genuine ornament from idle excrescence'.[9] This functionalist thinking is an example of a continuum stretching from the mid-nineteenth century to the Modern movement. The Modernists embraced functionalism in a more brutal form, but Pugin, in his time, created the essence of the true Modern building plan.

Certainly, to those who hold rigidly to the subtle strand of functionalism displayed in the thinking of Pugin – as distinct from adopting entirely Pugin's fervent view that only pure and authentic Gothic architecture should be the order of the day – an approach to development such as facadism (where we find, for instance, an elegant Georgian or a well-crafted Gothic revival facade preserved in front of a modern steel-framed concrete or brick box) must be something of an anathema. From such a perspective, a facade may be seen as mere decoration and the building nothing more than a piece of street architecture. The above stance is reflected by A. Dobby in his book *Conservation and Planning* (1978) where he expresses the view that facadism 'reduces buildings to mere elevations or self parodies' in the context of the pressures imposed by developers on local planning authorities, which can lead to the sacrifice of conservation principles. Such criticisms of facadism are founded on the idea that buildings should be designed three-dimensionally and not as exercises in street wallpapering or the creation of stage sets. However, as discussed above, this view has to be balanced with the long-established historical precedent for creating both individual buildings and street frontages primarily as architectural scenery.

A contemporary of Pugin was John Ruskin, who merged the design arguments upheld by Pugin with the social criticisms of people such as Thomas Carlyle. In his book *The Aesthetics of Architecture* (1979), Roger Scruton has made the following important observations about the form of functionalism espoused by Ruskin in relation to the way buildings are analysed:

> For example, there is a form of functionalism held up by Ruskin as the

Plate 5 Author's hand-drawn sketches based on A. W. Pugin's originals (1841).

> 'Lamp of Truth', a form which praises buildings for the structural honesty of their appearance. (On this view the Centre Pompidou deserves no praise, since what is revealed is structurally inessential, while the Gothic cathedral has much to recommend it even in an age when it has lost its principal use.) To elevate this principle into a dogma would have surprising consequences, for example, that St Paul's Cathedral, which hides functional buttresses behind screens of stone, is of less merit than the air terminal at Heathrow, or that the RAC Club in Pall Mall, which disguises steel and concrete behind a classical facade, some of the columns being mere hollow channels through which drains are conducted, is so debased as scarcely to deserve the name of architecture.
>
> (Scruton 1979)

Certainly, under Ruskin's 'Lamp of Truth', facadism can be seen as a dishonest and distasteful concept, but following Scruton's line of thinking, many buildings widely revered as great architectural treasures, if analysed in the same light, could similarly be condemned because of a lack of structural honesty. Therefore, unconditional rejection of facadism in any of its manifestations, purely on such a basis, is surely unreasonable. There are other qualities and characteristics of buildings to weigh in the equation. This is particularly so in cases where the facade is clearly the only part of a building with any architectural merit. For instance, a design philosophy much in vogue in Glasgow during the nineteenth century was the construction of commercial buildings in the City Centre, where enormous trouble and expense were lavished on the facade but the interiors provide a strong contrast, being utilitarian in the extreme. Many of these facades have their own intrinsic value for a variety of reasons. For example: some display great craftsmanship, or an elegance of proportion; others create a sense of drama within the townscape, and others have a certain rarity value. The last is illustrated by the Ca'D'Oro building at the junction of Gordon Street and Union Street, in the heart of Glasgow's City Centre (see Plate Section, x). Built in 1872 as a furniture warehouse, its exterior was modelled by architect John Honeyman on a fifteenth-century Venetian building of the same name. It is perhaps the most exquisite example of a small number of part stone, part cast-iron fronted buildings in the centre of the city. In the 1920s the building acquired a most unfortunate concrete mansard roof, as part of a conversion scheme into a restaurant and ballroom, which gave the building a top-heavy appearance and was quite at variance with the elegant original exterior. In the late 1980s, the opportunity was taken to remodel the basic and defunct interior to provide good quality retail and office space – wherein some 1920s stained glass was retained – remove the 1920s mansard and restore the facade, which was also extended to replace an adjoining ferro-concrete building. In this case a rare and important facade has been given renewed life by a radical internal remodelling of the building.

Those who hold rigidly to the 'Pugin line' should recall David Watkin's highly influential book *Morality and Architecture* (1977), in which he reveals the somewhat ambiguous nature of the arguments put forward by Pugin and championed by Ruskin. As Glancey recognises, Watkin undermines the notion that 'a particular approach to architecture could be morally right'.[10]

While considering facadism in the light of criticism that it is a distasteful concept because of its structural and functional dishonesty, it is interesting to note the approach to early steel-framed and high-rise buildings in America and Europe, and such buildings even as late as the 1930s in Britain. Many of these buildings now command immense popular appeal, whereas the much more functionally honest Brutalist blocks of the Modern era have generated bitter animosity from the public. The association between high-rise and steel-framed structures and the Modern movement is, of course, very close. However, the technique of using a load-bearing metal frame, which allowed the use of structurally independent external walling, and created the potential for an architecture of great constructional and functional honesty – one of the great tenets of Modernism – pre-dated the advent of the mainstream Modern movement, emerging in America in the 1880s. With a few exceptions, the development of the revolutionary structure was not at first followed by changes in style, which remained largely eclectic. Wall panels were used to hide the steel-frame structure rather than to emphasise it, and tended to be decorative: for example, in the case of the Woolworth Building in New York by Cass Gilbert (1913), which has 52 storeys of Gothic-style external detailing. Other popular motifs were classical columns and entablatures. The Ritz Hotel in London is an example of typical British steel-framed buildings of the early 1900s, where the cladding is used to mask the structure. Even during the 1930s, with the influence of Modernism growing in Britain, and the development of more plain elevations and repetitive fenestration, the tendency was still not to use the structure as part of the architectural style. A typical example is provided by the University of London Senate House (1933–7) by Charles Holden. Current cases of facadism, involving the retention of a facade, or the construction of a new facade in a traditional style, to hide a steel-frame structure, are in essence a reflection of such an approach.

Taking a moral stance in upholding or condemning particular design approaches to buildings is fraught with perils because the world of architecture raises many contradictions. The debate on facadism includes the following interesting dichotomy. The time-honoured tradition of refacing buildings is, in retrospect, seen as acceptable, or at least does not attract widespread condemnation, whereas the current process of retaining a historic facade, or creating a facsimile, in front of a modern building, attracts considerable criticism that it is a dishonest and distasteful practice. Leaving aside conservation issues, which are dealt with later in this book, there seems to be little justification, in architectural terms, for arguing that one practice is, in principle, any more honest, tasteful or acceptable than the other. To set this issue in context, it should not be forgotten that a large proportion of the buildings nowadays receiving statutory protection in various countries for their architectural or historical interest have been changed considerably from their original state. Some will have been given new facades in the past, and others will have been greatly adapted inside. Admittedly, facadism constitutes a drastic form of remodelling buildings, but it can be seen as simply a part of the ever-constant process of adaptation of the built environment, and an approach which has been in vogue over the last fifteen years, just as the refronting of buildings has been fashionable at times in the past.

Some opponents claim that because it is a backward-glancing concept, facadism is denying an architectural continuum and the emergence of new progressive styles. There may be some validity in this argument, because in having to retain an existing facade, or ape a particular style of facade, architects are denied the opportunity of giving an outward expression to buildings, which otherwise might be chosen. In this respect creativity is stifled. However, there can be little doubt that facadism, at least in Britain, is a product of the period since the mid-1970s, and indicative of the spirit of the age, which has seen public and professional opinion on architecture and town planning widely seeking solace and inspiration in the past. This is nothing new. A review of architectural history clearly shows that a reference to the past has frequently been a fundamental consideration in the formation of new compositions. For example, Palladio's reworking of the established Renaissance styles into new ideas and Pugin's revival of Gothic architecture were mentioned above.

The highly respected style of Art Nouveau should not be forgotten in this respect. As Doreen Yarwood indicates in her *Encyclopaedia of Architecture* (1985), Art Nouveau was 'based upon backward-glancing, and upon a return to craftsmanship' and the style itself was 'concerned more with surface decoration than with plastic structure'. An extension of the ideas of William Morris and John Ruskin, Art Nouveau was an 'aesthetic, romantic, ephemeral movement' (Yarwood 1985), which manifested itself in most European countries between the late 1880s and the First World War. Historically, it is important in architecture as an 'early attempt to break away from eclecticism' (Yarwood 1985), although this was not entirely successful. The movement, which focused originally on decoration rather than architecture, sought a return to the craftsmanship of the past, and iron, coloured glass and stucco were particularly favoured materials. The movement provided a route for architects seeking new styles, but who could not come to terms with the increasing emphasis on industrialisation and the pressures of the technological age. An extreme example of Art Nouveau is provided by Antonio Gaudi's curvaceous Casa Mila in Barcelona (1905).

Although acknowledging that facadism is not a stylistic concept, there are certain conceptual parallels which may be drawn with Art Nouveau. There is, for instance, the backward-glancing dimension, the finding of a solution in the past to the question of the outward appearance of new development. Also, facadism can be seen as a form of surface decoration to new buildings. A replica brick facade, for example, is likely to be only a 4-inch skin in front of a steel-frame structure. Another parallel is that facadism reflects a yearning for a return to buildings where craftsmanship and ornamentation are unashamedly displayed on the exterior. Many of the facades preserved nowadays as part of redevelopment schemes display examples of wonderful individual craftsmanship not available today. Interestingly, Art Nouveau emerged just as the Modern era was dawning and provided a diversion for architects who resisted the complete break with the past. At the setting of the Modernist sun, concepts such as facadism emerged, also out of a reaction against many of the buildings of the Modern movement, with their functional emphasis on technology and austere lack of decoration and individual craftsmanship. In a sense, facadism may be seen as a form of escapism from the perceived horrors of Brutalism.

Pulling the above threads of thought together, it is clear that there are conceptual links between what is referred to as facadism today and certain approaches to building design in past eras, and if justification for facadism is being sought in architectural terms, it is to be found on the basis of historical precedent. Clearly, an architectural continuum stretching from the Renaissance to the present shows that some schools of thought have had no difficulty, philosophically speaking, in divorcing the interior and exterior of buildings, or for treating the facades of buildings and street scenes as a fundamental design priority. Also, the built facade has, in its own right, obtained intrinsic value as an architectural element. Facadism can be seen as a continuation of these conceptual threads. As one of a number of particular responses to the need for urban change, facadism is in architectural terms a process of fusing the best elements of some old buildings, i.e. the facade, with new updated fabric behind. It is also a process concerned with contextualising new development with its surroundings. These concepts are not original but have been handed on by past generations of architects and town planners.

MODERN

So far in this chapter, facadism has been explored in the light of ideas linked to the architecture of pre-modern times. Now the focus moves to the ideologies embodied in Modernism and Functionalism, which have had such a dramatic influence on the built environment throughout the world. In the context of this book, it is important to examine facadism in relation to such ideologies, which sit in stark contrast to concepts such as facadism and represent a fundamentally different approach to dealing with the built environment. An examination of the dichotomy embraces not only stylistic matters, but also social, economic and political issues and the vexed question of morality in architecture.

Modernism represented much more than a stylistic break from the past. It constituted a moral crusade which required that all art should reflect the 'spirit' of the industrial age – the collective consciousness – and it was expected that the new architecture would meet the functional demands of twentieth-century life in an appropriate form. The Modern movement represented an extremely powerful international consensus which retained a significant momentum for about seventy years, although, in Britain, its impact was not truly felt until after 1945. Even then the pragmatic British nature was evident in signs of stylistic shifts in response to changing needs and fashions. However, an accepted corpus of architectural ideas remained intact, particularly an abhorrence of decoration for its own sake.

Although the period since the 1970s in Britain and Europe has not seen the like of the powerful international movement devoted to the dogmatic principles of Modernism, there has been a growing consensus in favour of the process of urban conservation, greater contextualism in new development, and the art of townscape has been rediscovered. Architectural history has been re-examined once more. This has been, in part, a reaction to a discredited Modernism, for many of the buildings created in the name of this dogma were merely pale images of the highly influential buildings of Mies van der Rohe or poor quality, watered-down versions of the early heroic work of Le Corbusier. The failure of many architects to live up to the exacting requirements of Modernist ideology resulted among the British public in a deep distrust of the architectural and town planning professions, and a turning once again to the past for design solutions. It could be said that a new moral crusade has been born in the form of the conservation movement, whose activists argue that they are fighting to save our architectural heritage from destruction, not only for the benefit of current generations but also for our successors. Some would argue that facadism is a part of this campaign, although, equally, others would deny its validity as a form of urban conservation (see Chapter 6, 'Facadism and Urban Conservation').

Facadism has undeniably played a significant role in many British towns and cities since the mid-1970s in allowing urban areas to adapt but, at the same time, retaining familiar street scenes which during the Modern era were being swept away. The prevailing ethos of the current age, if this can truly be identified, would be rooted in nostalgia and a desire to find reassurance, guidance and inspiration in tradition. Significantly, many manifestations of what has come to be known as facadism emerged at a time of considerable economic and political flux in Britain during the 1970s which corresponded with the dawning of the postmodern age. A period of considerable state intervention, which was in British terms relatively collectivist and had seen Modernism take root, was giving way to the era of the individual and the coming of the free market and monetarist ideology of Thatcherism. An eclectic explosion of taste ensued.

To understand the essence of Modernism and fully explore the contrast between its main principles and the concept of facadism, it is essential to consider the historical context within which the Modern movement evolved and eventually receded to give way to the emergence of another era, which has come to be known as postmodern. Therefore, a brief contextual background is provided below, which includes an identification of key characters or 'schools of thought' in architectural and town planning history. Clearly, an important aspect of the contrast being explored is the extent to which developments in architectural thinking have intermeshed closely with social, economic and political change and to highlight this I have focused primarily on post-1945 Britain.

To trace the first signs of the coming of Modern architecture, however, the search must begin in America during the 1880s with the work of Lois Henry Sullivan and Frank Lloyd Wright. Sullivan pioneered large functional commercial buildings, utilising the latest structural techniques, although his designs still remained decorative. Wright was to become a world architectural leader of the Modern movement in a career spanning seventy years. Before the First World War he designed mainly rural houses which reflected his strong view that dwellings should be an integral part of the landscape. Wright's 'prairie houses' were low buildings with gentle roof profiles possessing a significant plainness and clean lines, although traditional materials were used to assist the blending of the buildings into the rural setting. In his second design phase, Wright tended to abandon traditional materials in favour of concrete. In his house 'Falling Water' he dramatically explored the structural possibilities of concrete by cantilevering over a waterfall.

In Europe, Modern architecture began to emerge slowly after the First World War out of a deep and widely held desire for a wholly new architecture not based on past styles. Even before the turn of the century, the stirrings of a reaction to the elaborate decoration of mid-nineteenth-century architecture were evident in some designs, which, although still partly eclectic and using traditional materials, displayed cleaner, more simple lines and a minimum of ornamentation. A more fundamental change was evident during the early years of the twentieth century in the futuristic ideas of the Italian, Antonio Sant'Elia, and Tony Garnier in France, which pointed towards a whole new creed and the moral crusade of Modernism. Sant'Elia experimented with the romantic aspects of technology in his plans for cities, one of which – Città Nuova – was exhibited in Milan in 1914. His plans – which featured skyscrapers, multi-level roadways and pedestrian precincts – expressed an ideal socialist society, an important facet of Modernist ideology, and significantly influenced architecture in the 1920s. Garnier's idealised plans – for example, Cité Industrielle – featured illustrations indicating an extensive use of concrete in buildings, flat roofs and claddings of glass and steel.

There is an interesting political dimension to the contrast between Modernist architectural dogma and facadism. The former was inextricably linked to the socialist ideal, whereas facadism has emerged as an increasingly popular development option during a period when the socialist order in Europe has been eroded, to be replaced by more capitalist-oriented systems. In Britain, since the mid-1970s, the influence of the public sector over the development process has been reduced, while private interests have become more powerful. Could it be that concepts such as facadism reflect the architectural pluralism of the period that has replaced the Modernists' standardised functional design solutions?

The far-reaching experiments being carried out during the early part of the twentieth century into the use of pre-stressed and reinforced concrete fuelled the creative imaginations of talented architects and engineers and enabled Modernist architects to design and build radical new functional buildings. In his Church of Notre Dame, Le Raincy, near Paris (1923), Auguste Perret illustrated the structural potential of this material by carrying the wide-span roof on slender reeded columns, allowing the provision of non-load-bearing wall panels largely filled with coloured glass.

During its early years Modernism embraced 'constructivist' ideas – an abstract art movement which emerged in Moscow after the 1917 Revolution – wherein the structure of buildings was emphasised and decorations were regarded as trivia to be discarded. A closely allied and much more influential concept of the 1920s, and one which remained as an important tenet of architectural design throughout the Modern era, was Functionalism. This required that the form and construction of buildings was dictated by the need for proper and efficient function, and structure was to be 'visible and unashamed, not covered by a classical or other decorative facade' (Yarwood 1985). Taken to extremes, Functionalist dogma postulates that beautiful architecture is the inevitable result of designing buildings which function efficiently for a required purpose. Clearly, Functionalists would be unable to accommodate facadism within their views of how the built environment should be remodelled.

One of the most important and influential schools of Modern art of the twentieth century was the Bauhaus. Founded in 1919 at Weimar in Germany, after the amalgamation of two art schools, the Staatliches Bauhaus became world famous under Walter Gropius, and attracted architects, artists and students from all parts of Europe who were seeking opportunites to break free from established medieval or classical designs in all the visual arts. Gropius wished all his students, when qualified, to be able to understand and supervise both the aesthetic and constructional issues in their work. Not a new idea, this was practised by Renaissance and Baroque architects, and, for example, Adam and Kent during the eighteenth century, but such an approach had been lost during the nineteenth century. Gropius and his followers sought to approach building in a collaborative manner. Although the Bauhaus had this significant link with the past, its design concepts were forward looking. Gropius provided a synthesis of traditional concepts and the developing Modernism. During the 1920s, typical Bauhaus design became highly industrialised and functional, the resulting buildings possessing a stark, studied plainness, stripped of any gaudy decoration. Political changes in 1924 forced the relocation of the Bauhaus to Dessau, where Gropius's purpose-built school became a prototype for the new architecture. Even when the coming of National Socialism forced the final closure of the Bauhaus in 1932, when it was under the leadership of the influential practitioner of Modern architecture, Mies van der Rohe, its influence lived on well into the second half of the century. The design concepts of the Bauhaus project, which sought clarity and order of form, an honesty in the use of materials and dismissed eclecticism as inappropriate, provide an extreme contrast to facadism, where the combination of old and new fabric often results in disordered form and disparate visual elements and where the facade is used to screen rather than project the building behind.

Just after the First World War, the advanced and in some respects utopian concepts of Le Corbusier emerged as great influences on the development of Modernism, and indeed Le Corbusier became one of the great world leaders of the Modern movement. It is appropriate at this point to reflect briefly on the main elements of Le Corbusier's vision of the built environment, because it brings into sharp focus the contrast between the postmodern era of contextualism, adaptation of existing buildings, conservation and nostalgia to which facadism belongs, and the so-called 'brave new world' of Modernism.

Le Corbusier articulated his vision of Modern cities in his books *Vers une architecture* (1923; first published in Britain as *Towards a New Architecture*, 1927) and *La Ville radieuse* (1933; translated as *The Radiant City*, 1967). One of his pre-occupations and specialisms was the provision of mass housing within high-rise complexes of immense scale. Le Corbusier put great emphasis on providing large areas of communal open space around his buildings for the benefit of the occupants, and so each giant apartment block would have a spacious, verdant parkland setting. In his vision of 'A Contemporary City for Three Million People', exhibited in 1922, he planned to leave 85 per cent of the land as open space in the residential sectors. His building designs were characterised by simple structures without unnecessary and extraneous features, and the frequent use of flat roofs and plain concrete walls. The pilot system of freestanding concrete columns supporting the building above, as in L'Unité d'Habitation, Marseilles (1947–52) was his brainchild. As a housing concept, the Unité comprised a neighbourhood unit of apartments and appropriate commercial and communal facilities within a single block. Three more followed the Marseilles Unité, at Nantes (1952–7), West Berlin (1957) and Briey-la-Forêt (1960). Indicative of the Modern technological age, Le Corbusier saw the housing unit as 'a machine for living in', which offered a healthy, refreshing new style of living for working-class families who could live high above the fast-moving traffic, surrounded by fresh air and nature, and enjoy the benefits of balconies for exercising and sunbathing, safe places for children to play, rooftop swimming pools, and shops and cafés incorporated into each residential block.

The impact of Le Corbusier on the built environment has been immense, both in terms of the form and appearance of buildings and the way townscapes have been planned. In a British context, the provision of mass housing after the Second World War – in large high-rise blocks set in large areas of open space – clearly displays the Corbusian influence. For example, Powell and Moya's Churchill Gardens estate, Pimlico (1946–52) and the much admired early 1950s Roehampton estate sitting on the edge of Richmond Park. Also, the use of raw concrete in the Brutalist architecture of the Barbican estate in London, and Park Hill estate, Sheffield, echoed the Unités. Sadly, the British attempts to follow Le Corbusier's inspired ideas never realised the vision. A lack of amenities, poor quality detailing plus the inappropriate British climate conspired to turn an Arcadian dream into a nightmare. The horrors of the failed dream encouraged a reappraisal of how the urban built environment should be managed and remodelled, out of which ideas such as facadism surfaced.

While much Corbusian thinking appeared futuristic, there were some threads from the past woven into it. The importance attached by Le Corbusier to the relationship between the provision of open space, trees and plants and a successful living environment, linked him to Ebenezer Howard and the Garden City Movement.

However, his vision was very different to Howard's and had no place for the individual garden, which he considered too small to be of any real use. High-rise building provided the opportunity for the provision of park-style settings, extensive facilities for communal recreation and managed allotment gardens for those who desired them. Another link with the past is evident in Le Corbusier's Modulor, a system of proportioning built form, which was related to classical orders. Also, Le Corbusier claimed that his L'Unité d'Habitation concept had been influenced by a Carthusian monastery at Emna in Italy. It seems that even in the midst of such a powerful will to turn away from the past as was obviously the case during the Modern era, the complete severance of historical strands was never completely achieved. Referring to the 1960s in Britain, Jonathan Glancey notes in *New British Architecture* (1989): 'the work of the best mainstream architects of the mid-1960s was a collage of images and ideas drawn from such diverse sources as Inca villages and the late houses of Le Corbusier'.

Such considerations undermine the Modernist single strand theory and lend weight to the views of Charles Jencks expressed in *Modern Movements in Architecture* (1973), in which a series of discontinuous architectural movements and the plurality of approaches are postulated. During the Modern era, architectural theory became historicist or 'motivated by arguments which contend that there is one inevitable line of development' (Jencks 1973). Technological and structural determinism was the order of the day. However, as Jencks points out, advocates of a single strand or *Zeitgeist* theory place limits on the diversity of live traditions and look only for evidence to substantiate such theory while denying inconsistencies. Plurality is coalesced into an integrated movement. Yet the Modern movement did represent a remarkable international consensus based on an ideological and stylistic break with the past. Within such a consensus a concept such as facadism had no place, but from within the broad Modern church, which has continued and evolved throughout the postmodern era, facadism might appear less of an alien concept when we consider Punk and New Wave creations with their diverse mix of elements.

It is instructive to consider the social and economic climate within which Modernism evolved and became an approach to urban development rather than the traditions of the past. A number of factors provided the catalyst for its development. In Europe, there was an acute need to create buildings more quickly during the nineteenth century to cope with the effects of the Industrial Revolution and a massive population explosion. Later, the destruction caused by two world wars created vast rebuilding programmes. As a result, technical advances were hastened to allow the cheaper mass-production of large-scale buildings, which inevitably sounded the death knell of individual architectural craftsmanship. Internationally, Modern architecture utilised similar methods of construction, steel-frame structures being generally used in countries such as the USA, Germany and Britain, where steel was readily available. Elsewhere, as in Spain and Italy, architects relied more on ferro-concrete.

In Britain, the main thrust of Modernism was not felt until the late 1940s and 1950s when the post-war economic, social and political climate provided ideal conditions for it to take root and thrive. The ravages of war, and the harsh difficulties of high unemployment, the Depression and the worst effects of the industrial revolution in pre-war times, had created the necessary consensus within society to build a new Britain and break away from the past. The war had also united the classes and fostered a faith in the ability of the state to deal fairly with rich and poor elements of society in a time of crisis. Even though there must have been immense respect for Churchill's wartime leadership qualities, a Socialist government, under Clement Attlee, swept to power in 1945. This government was faced with a massive rebuilding programme and the state took on the mantle of architectural patron – a role which was to last until the economic crisis in 1973 – to provide, for example, new towns, vast numbers of new houses, schools, hospitals, railway stations, libraries, courts and government offices. Clearly, this provided tremendous opportunities for the emerging young Modern architects, and many became employed in the public sector. The situation of the time is summed up well by Jonathan Glancey:

> Young architects, steeped in Le Corbusier's ideas and a romantic vision of socialism, believed that architecture could start again from scratch, free from the onus of history and precedent. German bombs had already begun to start the process: it was time to rebuild an England without slums, class warfare and the worst effects of the industrial revolution.
>
> (Glancey 1989)

The Modern movement may have attempted to divorce itself completely from the past, but it is interesting to note that a century after Pugin's campaign to establish Gothic revival as a serious style, Nikolaus Pevsner was taking up a similar argument against the immorality of sham materials and technique in architecture. With a passion akin to Pugin's advocacy of pure Gothic architecture as the only morally correct style, Pevsner championed the cause of International Modernism as the 'only true creed'.[11] In the thinking of both Pugin and Pevsner, architecture is not just a style but 'a rational way of building evolved inevitably in response to the needs of what society really is or ought to be, and to question its forms is certainly anti-social and probably immoral' (Watkin 1977).

There is a significant link between Pugin and Modernist thinking but also a striking contrast in terms of an attitude towards tradition. Consider the following:

> Pugin's great concern was to model the plan of a house not on a formulated principle, as with the ideal Palladian villa, but on strictly empirical and pragmatic grounds. By building walls up from rooms shaped and sized according to their function, Pugin created the essence of the true Modern Movement plan. Yet his buildings abounded with historic reference and traditional craftsmanship.
>
> (Glancey 1989)

Within the ethos of Modernism, historic references and traditional craftsmanship were unwelcome. Pevsner reflected the antipathy within the Modern movement to the past, by his judgemental use of the term 'historicism' in relation to the influence of past styles in new architecture. The new architecture was to be integral, not to contemporary society, but to the idea of a socialist industrialism to be realised in the future. Faith rested on the emerging new technology, and individualism was to be subjected to one organic consciousness, the *Zeitgeist*, inherent in a cohesive collectivist society. The *Zeitgeist* was all-pervasive and thus architecture was required to reflect the 'spirit of the age', the economic structure and prevailing theology. There was also an evolutionary assumption that in each epoch a new *Zeitgeist* is 'struggling to be born' (Watkin 1977) and, as a result, 'the cultural, religious, moral and political patterns of previous epochs are rendered obsolete' (Watkin 1977). In reflecting the above ideals, architecture was to be distinguished by honesty, truth and anonymity in its form, design and appearance. The exterior of a building was to be a true reflection of the interior and materials were to be used without pretence.

> Modern architecture was as much a moral crusade as a way of building. It was seen as morally good as it told no lies. The elevations of a Modern building were drawn straight from its plan. A Modern architect would never, so the theory went, deceive his public as Sir Christopher Wren (1632–1723) had done for example at St Paul's Cathedral, where high Baroque walls hid a medieval Gothic plan, a narrow nave and flying buttresses. Modern architecture was morally right because it used materials honestly; no self-respecting Modern architect would pretend that one material was another. At the same time it was right because it sought to satisfy real, rather than imaginary, needs.
>
> (Glancey 1989)

The term 'facadism' emerged during the nadir of the Modern movement in Britain, when social attitudes towards the built environment were changing dramatically. During the mid-1970s, a sense of public outrage and revulsion emerged against many of the austere, anonymous and brutal forms of the Modern movement, which sought and achieved visual conflict with existing townscapes and turned many urban areas into concrete wastelands. Many of the buildings proved totally unsuitable for their proposed uses – such as high-rise housing accommodation lacking the necessary facilities to cater for families with children, and school buildings or offices which were cold in winter and too hot in summer. The exacting requirements of Modernist ideology – wherein the contradictions of freedom from tradition, on the one hand, and the constraints of having to design in terms of functional requirements, on the other, had to be overcome – were beyond the ability of many architects. Such demands required the touch of artistic genius, as shown by Le Corbusier, to be successful.

In the post-war years, the newly created planning system had been seen as the mechanism for creating a glorious new environment, and a number of different Labour administrations attempted to deal with the question of 'betterment' and development value tied up in land. Successive governments had exercised considerable architectural patronage since the war, but the socialist view was that by taxing development value there would be far greater scope for the state to take a much more pro-active development role. However, while the succeeding Tory governments were more interventionist than their more recent counterparts, they could not come to terms with such socialist policy and the legislation was repealed. Consequently, the necessary public investment to truly achieve the utopian post-war vision of a new Britain was not available.

Added to the poor quality of much of the architecture of the time, the failure of the planning system to live up to the public's expectations created enormous disillusionment. The architectural and town planning professions were perceived as arrogant, out of touch with the people and part of a corrupt conspiracy with the state and property developers, whose only interest was the voracious pursuit of vast profits. Buildings such as Richard Seifert's Centre Point, commissioned by property developer Harry Hyams and which lay vacant for many years, symbolised, in the public's eyes, the relentless greed of speculative developers encouraged by Harold Wilson's Labour Government. The jailing of John Poulson, the Newcastle city architect, in 1973 brought feelings about corruption to a head. Ironically, the Modern movement had sought to create an architecture which was truly popular in nature, but only succeeded in alienating the public, through a dogmatic adherence to a creed which often took little account of the needs or opinions of people who had to look at and use the buildings created.

As an antidote to the horrors of Modernism, the past was brought back into focus once more and traditional architectural styles and materials, craftsmanship and ornamentation returned to popularity. With the planning and architectural professions on the defensive, and public opinion in favour of reviving the past, the ground was prepared and fertile for the growth of the conservation movement and concepts such as facadism. The comprehensive redevelopment areas of local authority development plans of the 1950s and 1960s were gradually eradicated and urban design became a more incremental process of stitching the built fabric back together after the scars of brutalist development. The Modern idiom of bold freestanding buildings making their own statements, unconcerned about their context, was replaced by a spirit of 'urbanism'. Sensitive infill development, the retention of streetscapes through the preservation or rebuilding of facades and the re-use of old buildings have become key aspects of the process. Postmodernism is an era of compromise and contextualism when architectural metaphors are mixed into diverse and interesting cocktails as, for example, in James Stirling's Clore extension to the Tate Gallery.

The period since the mid-1970s has seen an upsurge of both professional and popular interest in conservation and, for many people, the cause of protecting Britain's architectural and historical heritage has been akin to a crusade. The campaigns have had considerable success, with the influence of conservation interests in the form of amenity societies and other pressure groups being significant in decision-making processes and the conservation dimension becoming an established element of planning legislation. The emergence of the practice of keeping old facades or creating facsimiles has been consistent with the upsurge of popular support for conservation, because it at least has enabled familiar street scenes to be preserved intact while, at the same time, allowing pressures for up-to-date accommodation to be realised.

Another instrumental factor which led, certainly in the 1970s and 1980s, to a considerable number of cases of facadism, was that many of the buildings surviving the wholesale clearance and redevelopment schemes of the Modern era had suffered years of neglect and were in a very poor state of repair. A survey in Bristol in 1977, for example, revealed that within the central area of the city there were 479 historic buildings in a derelict condition.[12] Throughout the country, it was found that these historic buildings were often beyond economic restoration and conversion to present-day standards and, in some cases, the facade was the only part of the building which was structurally sound enough to be saved. Even then there have been occasions when facades have collapsed during the process of trying to preserve them *in situ*. Frequently, decisions were taken to preserve or rebuild facades in front of modern accommodation, rather than allow the creation of a completely new building, such was the extent to which Britain had been gripped by a lack of confidence in the architectural profession following the 1960s and 1970s.

As well as the shift in social attitudes towards the built environment, significant changes were also occurring in the economy and the political climate. The economic crisis of 1973 brought an end to the extensive state patronage, which had created so much architectural work since the war and fostered large-scale redevelopment projects such as public housing schemes polemically inspired by Le Corbusier and his followers. The private sector had now become the main source of major development projects and architects were obliged to negotiate with individual clients, who had more opportunities to influence architectural form and style. Also, the view that high-rise and system building was not the economic miracle it had once seemed gained popularity, and that the structural implications of many experimental Modern buildings, erected hurriedly, had not been well thought out, with costly consequences both socially and economically. Some cities introduced embargoes on further large-scale office building, as a result of high vacancy rates caused by the economic difficulties in the mid-1970s.

In this mixed climate of having greater involvement as patrons yet being affected by the economic difficulties, private developers, particularly those involved in the office sector, turned to concepts such as facadism as a possible solution to securing planning approvals. The retention of a facade, or creation of a facsimile, offered a lever – a conservation carrot – to the planning authority in return for new office space behind the facade. Faced with the likelihood at the time that many historic buildings would fall down in the near future because of their critical condition – planning authorities such as Bristol did not have the resources to restore the numbers of derelict listed buildings involved – the developers' proposals to save or replace a historic facade often appeared to be the only practical option. Delays in trying to negotiate full restoration or rebuilding schemes could have deterred developers or caused further irremediable damage to the building fabric. Facadism, therefore, became an important element in the transition from the era of Modernism to the present postmodern age, and continues to be a significant concept in urban remodelling.

It is clear that, in the light of Modernist ideology, facadism would be an anathema, an immoral and worthless approach to creating buildings. Such condemnation will now be analysed critically.

A fundamental objection would be that the exterior of the building is not a true reflection of the interior, that form does not follow function and the facade is simply used to hide the structure, rather than the structure being used as an architectural element. The result is fake, dishonest buildings, as in the case of new buildings erected behind preserved facades where the internal design of the building is likely to be influenced by the proportions of the facade as much as the function of the building. For instance, floor levels may be dictated by the window positions in a retained facade. Robert Venturi, the Philadelphian architect, in his highly influential book *Complexity and Contradiction in Architecture* (1966), pointed to the antipathy in Modern architecture towards 'the false front'.

Earlier in this chapter, the condemnation of facadism from the standpoint that it denies structural honesty in architecture was dealt with in the light of the pre-Modern views of Pugin and Ruskin. They upheld a subtle strand of functionalism and some of the points discussed in that context apply to a consideration of facadism in relation to the Modernist view on function, although the functionalism of the Modern architecture was much more rigorously applied, especially in terms of ornamentation. A counter-balancing view to the dogma that form must always follow function is evident in the words of Roger Scruton:

> function is not the only essential feature of a building; necessarily a building must display a form or pattern, and we may come to understand the pattern in contexts divorced from the given function. Consider, for example, Alberti's beautifully incrusted facade to the Florentine Church of S. Maria Novella. Many of the patterns there exhibited might have formed part of an ornamental cloth, and been found similarly beautiful. Moreover, to remove them from the facade – to present a bare face of serpentine or marble – would be to remove a major part of its charm, while preserving whatever deserves the name of an expression or revelation of its use.
>
> (Scruton 1979)

Clearly, there are many aspects of design and structure which contribute to an understanding and appreciation of buildings, and facades are highly significant in this respect. In its striving for complete honesty in design and expression, Modern architecture denied the importance of the facade and its role in communicating meaning to the onlooker.

> It is surely not an accident that we like to see our buildings *stand* before us, not if we derive our conception of unity and humanity in building from our own sense of self. Nor is it an accident that we like to see a street as formed from a series of facades, or that the Brutalist's 'street in the air' seems such an inadequate visual substitution for the back-to-back slums which preceded it.
>
> (Scruton 1979)

The proliferation of horizontal emphasis in much Modernist architecture seems, to Scruton, inimical to the outward projection of self into form and that 'In the upward movement of lines on a facade we feel the moral force of human posture' (Scruton 1979).

Facadism is a process which takes into account the unique significance of the facade as an individual design element – an idea that has a distinguished historical pedigree, as discussed earlier in this chapter – and allows streets to retain their series of facades and a relationship with humanity. In the sense that facadism appeals to the human desire to be able to relate to the exterior of buildings and to gaze upon works of high quality or beauty, it could be argued that it is more of a popular form of design than most Modern architecture. Scruton is in no doubt concerning the importance of the facade, and points to the Modernist folly of relying on functional architecture to create meaningful buildings, arguing as follows, in powerful language:

> A facade gives a building some independent stance, a meaning that it is *not* merely borrowed from the activity which it contains. The visible commotion of its inmates no more gives humanity to a building than the wriggle of worms gives vitality to a corpse. The soul of a thing is of its essence; it cannot be simulated, borrowed or stolen from some foreign source. A building without a facade is not just a building without a face – it is a building without expression, and hence a building without life.
>
> (Scruton 1979)

The Modernist dogma that the exterior of buildings should truly reflect their interiors also denies the multiplicity of roles that buildings can perform. For instance, an insistence that a building, which of necessity requires a sober interior to accommodate a particular use, must not have a different outward character and perhaps entertain or create humour on the street, is no more justified than refusing a brilliant comedian an opportunity to perform because he or she is inwardly a shy and retiring character. I have encountered the argument that facadism results in nothing more than the charade of 'Disneyland' and stage-set architecture lacking reality and truth. However, does this constitute valid grounds for objection? Is not one of the roles of buildings to provide an enjoyable setting for life's activities? If the preservation of much loved facades or creation of facsimiles enables this, then such a process rises above the criticism of a particular dogma.

Before leaving the debate about the conceptual separation of exterior and interior, the following thesis should be noted. In practice a building has three distinct functions. The outer skin, the facade, presents a face to the outside world and is very much a public matter. The interior provides the setting for the users of the building and tends to be a private rather than a public concern. The third function concerns the structure which holds the building together. Such a thesis suggests that the exterior of a building need not necessarily reflect the interior as the functions of both are quite different. Venturi referred to the role of the interior and its conceptual separation from the exterior as follows: 'The essential purpose of the interiors of buildings is to enclose rather than direct space, and to separate the inside from the outside' (Venturi 1966).

There would seem to be every justification for cases of facadism which reflect both the public's wishes to retain or create an environmentally important street scene, and the particular requirements of the private occupiers relating to the interior. Facadism may be seen as a marriage of convenience, but not one that necessarily ends in divorce. The partners may survive together, fulfilling separate roles, to forge a meaningful relationship.

The word 'facadism', implies a preoccupation with the exterior of buildings and 'the street', but, in reality, there is another aspect of the concept which should be considered. Facadism, rather than simply allowing the preservation or enhancement of a streetscape or the projection of a particular image for a developer or an occupier, also releases the opportunity – through the removal of old fabric with its inherent constraints on how buildings can be used or adapted – to create interiors which are far more interesting, exciting, adaptable, aesthetically pleasing, comfortable or appropriate than the interiors of some older buildings. It could be said that facadism encourages a healthy reassessment of how interiors work and are designed. A fitting example is to be found at 'Atrium Court' in the centre of Glasgow, where an early twentieth-century building fronting West Campbell Street, Waterloo Street and Wellington Street was adapted in 1989 from a former GPO parcels sorting office into prestigious offices now occupied by the Glasgow Development Agency as its headquarters. The heroic facade was retained and the semi-industrial interior demolished, to be replaced by attractive modern offices wrapped around a soaring atrium (see Plate Section, i, ii, iii).

The radical interior redesign opportunities presented by facadism – as opposed to the process of simply adapting an old building in its existing form, with all the constraints that this places on architects such as floor heights, staircase positions, etc. – may also enable a greater diversity in the use of buildings. In some cases, greater public accessibility may result from the adaptation of buildings and, where this occurs, the public has an opportunity to experience interior space as well as the street scenes. For instance, this has occurred in Bristol City Docks, where two former warehouses have been transformed in different ways. In one case, a Bristol Byzantine-style facade stands in front of a new Squash Centre. In the other, the former Bush Warehouse was gutted inside to create the Arnolfini Art Centre, containing galleries, a cinema, bar, restaurant and shop.

Derogatory references to façodomy within the Modernist design community in the United States, relating to the practice of preserving facades in front of new buildings, suggest a debased and impure form of building design. The emotive term surfaced in the debate surrounding a contentious development on Red Lion Row, Washington DC, where the frontage of a nineteenth-century terrace was retained and linked into an eleven-storey office block erected behind. This particular case which results in the bizarre and seemingly meaningless union of the most unlikely architectural bedfellows, where little attempt has been made at arriving at a cohesive composition, leaves the impression that the Modernists may have a valid point! The preserved terrace frontage lies totally dominated and struggling to express its identity at the feet of the menacing Modern giant thrusting upwards behind.

From a Modernist perspective, facadism would be condemned as a backward glancing and eclectic process. Although it has such characteristics, facadism is also a forward-looking concept in that it involves adapting and updating urban fabric to meet present-day demands and needs. It should also be recognised that, even in the midst of vehement desires to turn away from history and tradition, a complete severance from the past was never truly achieved by the Modern movement. I referred earlier to the influence of the ancient Golden Section in Le Corbusier's ideas on proportion formulated in *Le Modulor.* In a British context, during the heyday of Modernism in the 1950s, the 'Festival of Britain' attempted to demonstrate a fusion of picturesque and whimsical Modern architecture with the traditional concepts of town planning. Many of the most talented Modern architects were Classicists at heart, 'fully in tune with history but unwilling to ape its forms' (Glancey 1989). Lubetkin combined ancient and Modern in his Highpoint flats, which were much admired by Le Corbusier. There, he chose to hold up the entrance canopy to Highpoint II with exact copies of the caryatids borrowed from the Erechtheum at the Acropolis in Athens (see Plate Section, xiii). It seems that the thirst to look to, or rework the past, can never be totally assuaged in architecture. Human beings have a basic need to refer to the past, to their roots, to find a sense of time and place, and in reality, facadism has become an important part of this process.

It can be argued that where existing buildings are obsolete or derelict, they should be replaced with completely new buildings which are of the present, and that propping up facades or creating facsimiles (and thereby preserving images of the past) is unjustifiable as it stands in the way of a constantly developing architectural continuum. There is a dominant historicist element in such an argument. The philosopher Karl Popper, in *The Poverty of Historicism* (1957), wrote: 'historicism claims that nothing is of greater moment than the emergence of a really new period', and later in the same book: 'Every version of historicism expresses the feeling of being swept into the future by irresistible forces ...'

Pevsner, the great advocate of International Modernism, uses the term 'historicism' in a generally condemnatory sense in referring to the aping of past styles in architecture. However, as David Watkin indicated, Pevsner's view is:

> historicist in precisely the Popperian sense. It is holistic and preoccupied with the future. He believes that art is and should be a product of the economic, social and political conditions under which it is created; he believes that there is such a thing as the essence of an age and that the common essence is more important than its individual manifestations; he has accordingly discerned the political and social norms of the twentieth century, and insists that art and architecture must be subordinated to them. Consequently he can and does insist that he knows what will and must be the architectural expression of the age.
>
> (Watkin 1977)

There is a poverty in such an analysis of the philosophy of architecture, in that it denies the value and richness of what has been produced in the past and its relevance – to bring pleasure, inspire, educate, or give a sense of wellbeing in a fast-changing world – to present and future generations. Condemnation of facadism from a Modernist perspective should, I believe, be seen in such a light.

POSTMODERN

In considering facadism from this perspective, I have taken postmodernism in its broadest sense. Since the ebbing away of the great wave of optimism and ideological motivation associated with the Modern movement, a diverse range of architectural styles and ideas has emerged including for example, the 'High Tech' or 'Late Modern' of Norman Foster and Richard Rogers, Nigel Coates's 'Industrial Baroque' or 'New Wave', the 'Neo-Classicism' of Quinlan Terry and Leon Krier, and the work of Terry Farrell and John Outram which is often specifically referred to as 'postmodern'. Early British examples of postmodern architecture were highly influenced by the Americans, witty and ironic but somewhat contrived, such as the clumsy games played at Classical revival with stuck-on or painted pilasters and pediments, or the Hollywood Strip mentality found in Farrell's 1982 headquarters for TV-am, where he reworked a former garage in Camden Town. This fragmentary cut-and-paste collage approach has roots in the mid-1960s thinking of Robert Venturi, whose highly influential book *Complexity and Contradiction in Architecture* (1966) represented the dawning of the postmodern age. However, as indicated above, the postmodern framework contains numerous strands of thinking. The continuing Modern church is nowadays far less idealistic and much more pragmatic. Coates's 'New Wave' work displays outrageous humour blended with well-crafted durable design and brings art, craft and fashion into architecture – once more echoing American Art Deco and the British Arts and Craft movement. Quinlan Terry has brought both traditional classical design and building methods to the fore once again, as in his eight-storey office development at Dufours Place, Soho, in 1982, which uses load-bearing brick. Terry Farrell has become greatly involved in large-scale urban renewal schemes and, as the chair of the Urban Design

Group, has focused on sensitively filling the 'gaps' in British cities. Farrell's 1987 design for a large office scheme straddling London Wall in the City of London can be seen as 'a significant step for Post-Modern respectability' (Glancey 1989).

Clearly, in the postmodern age, the past has become respectable once more and indeed highly influential, as witnessed by the involvement in the architectural and town planning arena of HRH The Prince of Wales. In his 1984 speech to the RIBA, the subsequent TV documentary *A Vision of Britain* (1988), and the book of the same title (1989), he pilloried the professions for:

> [the] wanton destruction which has taken place in this country in the name of progress; about the sheer, unadulterated ugliness and mediocrity of public and commercial buildings, and of housing estates, not to mention the dreariness and heartlessness of so much urban planning.

The Prince was taking issue with the Modernists' rejection of the past, and likened it to people losing their souls. His views found a captive and enthusiastic audience in the British public.

While there could be no place for it under Modernist ideology, facadism has become an established concept in the postmodern age and has continued since the mid-1970s to play a significant role in the process of urban change and remodelling. It sits easily within the often inclusive and contradictory world of postmodern architecture and design. It has, itself, such characteristics. The inclusion of traditional facades, which may have been designed in terms of, say, classical principles, on buildings that have totally different proportions and style inside is an obvious contradiction. For example, the retained Georgian facade at 1–6 Brunswick Square, Bristol, with its party quoins and false chimneys, contradicts the open-plan office building to which it is attached (see Plate Section, iv, v and xx) but allows the 1980s development to include a vital reference to the townscape where the office building fronts the formal Georgian square.

Robert Venturi propounded the validity of contradiction in architecture and the ability of the creative designer to combine fragmentary images into a unified whole. Although Venturi's views were not totally focused on the past, they recognised the relevance of using historical references such as Classicism and Art Deco as well as current images in new architecture: 'architecture is necessarily complex and contradictory in its very inclusion of the traditional Vitruvian elements of commodity, firmness and delight' (Venturi 1966).

In seeking to distinguish between the Modern ethic and Venturi's inspiration, Vincent Scully wrote in his introduction to Venturi's book:

> Le Corbusier's great teacher was the Greek temple, with its isolated body white and free in the landscape, its luminous austerities clear in the sun.
>
> Venturi's primary inspiration would seem to have come from the Greek

> temple's historical and archetypal opposite, the urban facades of Italy, with their endless adjustments to the counter-requirements of inside and outside and their inflection with all the business of everyday life: not primarily sculptural actors in vast landscapes but complex spatial containers and definers of streets and squares.
>
> (Scully, in Venturi 1966)

This analysis points both to a recognition of the distinctly different roles of the inside and outside of buildings and the involved nature of the relationships between buildings, the spaces and places separating them and the activities contained. Venturi takes fragments from other styles, ideologies and his own original images in forming his compositions. Relationships are compromised, but there is a recognition of and a respect for what already exists. Contrast this with the singularity of much Modern architecture, the almost suburban nature of its designs with buildings often separated from their neighbours by large tracts of open space, and the deliberate disrespect for existing traditional development.

There are parallels to be drawn between Venturi's views on the built form and the facadist approach. There is, first, the acceptance of a conceptual divorce between inside and outside. Second, facadism is a compromise solution to the problems of urban development which pays heed to what are often competing demands: that is, the importance of facades in the defining of streetscapes and formal townscape compositions such as terrace architecture and squares, against the demands for up-to-date and adaptable interiors.

Facadism is a fragmentary process in that it involves taking elements or fragments from the past (either preserved or facsimile facades, or as in the refronting of buildings, the retention of interiors) and bonding them to new fabric in the form of decorative features. This is comparable to the early postmodern Classical revivalist attempts where historic elements were painted or stuck onto the buildings.

Venturi's perspective of architectural form is pluralist and fragmentary. It attempts to embrace the complex interaction of different facets of human experience found in urban areas and to work the fragmentary elements into meaningful buildings: 'an architecture of complexity and contradiction has a special obligation towards the whole: its truth must be in its totality. It must embody the difficult unity of inclusion rather than the easy unity of exclusion' (Venturi 1966). In an obvious riposte to the Miesian view that 'Less is more', Venturi claims that 'More is not less'.[13]

If this notion of truth in the totality of architecture is applied in the wider context rather than simply the individual building to the process of urban renewal, where, for example, a part of a city centre is being remodelled, new development will need to relate to a complex system. It may be that the inclusion of particular historic references are essential in uniting the whole development with its context because of what already exists and is going on around – the townscape and uses contained. An appropriate approach may be to retain certain facades or create facsimiles and knit them into the

Plate 6 Detail on the facade of a former warehouse, Bathurst Basin, Bristol.

remodelled urban fabric, because they convey important truthful messages about the area's past, or enrich people's experiences, or function as vital landmarks. The Bristol Byzantine warehouse facade in Bristol's Bathurst Basin (see Plates 6 and 7) retained in front of a Squash Centre, which is knitted into a dockside inner city residential development, reminds us of the truth that the dock basin had an industrial past and an employment function, once important to the local economy which saw Bristol become such a wealthy city.

However, taken too far, facadism denies architects the opportunity of expressing, in an original and creative way, the complex and contradictory nature of both the new development and the surroundings. Retained, replicated or mimicked facades may perform valuable roles, as suggested above, but the decision about the outward image to be portrayed is taken out of architects' hands even before the design process begins, and the chance to make more relevant, challenging or stimulating statements is lost. For example, had it been decided to create an art and cultural centre in Paris within a building which outwardly mirrored the adjacent historic development – for instance, through the construction of facsimile facades evoking the traditional styles of surrounding buildings – Rogers and Piano would not have created the shockingly exciting Centre Pompidou which, even fifteen years after its construction, remains as popular as the nearby Eiffel Tower, showing that the building has much more than mere novelty value. This Pop fantasy exudes life and vigour and as you gaze from the surrounding historic streets the sight of people moving through the transparent tubes suggests blood corpuscles coursing through a body's exposed veins. To be truly inclusive, architecture needs to be free of dogmatic stricture. At times it will be right to reflect or revamp the past. The vogues of the present day should be reflected even in traditional settings and this may involve a reworking of past imagery. It is also important to accommodate the art of contrast, the creative imagination and the futuristic vision. Contextualism is only one of a number of relevant issues to be addressed by architects and town planners in decision-making, and there are numerous aspects to contextualising new development. There is a place in this plurality for facadism, just as there is for other approaches to creating new buildings which rely heavily on the past for a solution. However, there is the danger that a preoccupation with the past becomes debilitating and stifling to innovation.

Some further words of Robert Venturi lead to another point about facadism: 'A valid architecture evokes many levels of meaning and combination of focus: its space and its elements become readable and workable in several ways at once' (Venturi 1966). In the case of terrace infill, facadism is often difficult to identify immediately, especially by the casual observer. However, in the case of freestanding buildings, where just the front facade is retained or rebuilt, it is often obvious. The junction of old and new can be surprising, and sometimes discordant. This is not to say that such an approach cannot result in valid architecture. The building may convey a number of messages at the same time. For example, at Brunel House in Bristol (see Plate Section, vi and vii) Pope's retained Classical facade tells the onlooker that it is worthy of its place in the 1980s

Plate 7 Photograph of the facade of a former warehouse and the adjoining townscape, Bathurst Basin, Bristol.

redeveloped building in its own right. It also provides images of dignity and reassurance through traditional values to the visitor approaching the City Planning Department offices. The new exterior elements advise that this is up-to-date office accommodation. As you pass through one of the monumental arches of the facade, the theatrical outside piazza – with a backcloth of retained walls built into the adjacent hillside – and modern sculpture refer to the age when horse-drawn carriages brought travellers from Brunel's Great Western Railway station to his hotel, prior to departure by sea to America. In the 1960s it was proposed that the whole hotel should be demolished and replaced with a glass-box office building, which would have had a certain functional honesty and have lacked the ambiguity of the present building. However, a range of meanings and *foci* are evoked all at once by the present building, and for this reason it possesses a considerable richness not found in a great deal of British Modern architecture. On the question of ambiguity William Empson's observation that ambiguity 'may be used to convict a poet of holding muddled opinions rather than to praise the complexity of the order of his mind'[14] is, I feel, most perceptive.

I referred above to contextualising new development. Much of Robert Venturi's thinking in the mid-1960s signalled the emergence of contextualism, which has become such an important and influential idea during the postmodern era.

It could be argued that, as a concept, facadism should be considered in the same light as completely new buildings which amount to collages of architectural imagery, such as Stirling and Wilford's Staatsgalerie in Stuttgart where Classical, Modern and Pop elements are brilliantly fused into a complex and contradictory architectural composition. The retention of facades or creation of replicas, for instance, could be similarly incorporated into new buildings as pieces of historical imagery.

I referred above to the debilitating nature of a preoccupation with the past to the exclusion of a focus on the present and the future, with a consequent stifling of creative imagination. Robert Hewison's polemical *The Heritage Industry* (1987) points to the pervasiveness of the past and nostalgia in the British postmodern culture. He recognises that it is made up of 'contradictory elements', but, in condemnatory tone, argues as follows:

> The prefix suggests that it is whatever is happening to contemporary culture, now that the great wave of modernism, lasting roughly from Picasso's first cubist paintings to the dropping of the atomic bomb, is over. Postmodernism is modernism with the optimism taken out. The prefix also suggests that nothing has happened since. At worst, postmodernism is the complete rejection of the present that leads to King's Walden Bury and the lifestyle of the New Georgians. At best, radical postmodernists recognise that we are living in a broken culture, and must make the fragments into a new pattern, not shore them up against our ruin.
>
> (Hewison 1987)

In the same work Hewison considers that postmodernism betrays 'an unhealthy dependence on the past'.

In Britain, the arrival of the enterprise economy and 'Thatcherism' at the end of the 1970s created the climate for the expression of a much greater diversity of individual taste and an explosion of architectural styles, many of which looked back in time for inspiration. Margaret Thatcher in 1983 turned the spotlight fully onto the past in calling for a return to 'Victorian values' and symbolically underpinned the architectural and town planning re-examination of the past which was already well under way. Irrespective of its validity in terms of the principles of urban conservation, facadism is in tune with the current British culture of nostalgia, and widespread uncertainties about future social, economic and political issues, which lead to seeking security in the familiar past.

Since the mid-1970s popular opinion on architecture and town planning has increasingly reflected the 'green' age of conservation and 'contextualism'. It seems somewhat ironic now that the architecture of the Modern movement was also intended by its pioneers to be truly popular.

In many cities in Britain and Europe, efforts have been made by central and local governments to either remove or adapt the legacy of Modernism in the form of poorly constructed and ill-thought-out high-rise buildings. Some have been demolished, others reduced in height and remodelled. In some cases, tower blocks have been refurbished and upgraded inside and turned into luxury flats. Many have been given a facelift through a range of treatments akin to the practice of refacing more traditional buildings. This is in essence a form of facadism, which the Modern and Brutalist architects who designed the buildings would have, at the time of their construction, found unthinkable. For instance, some concrete exteriors have been painted, others have been covered with decorative sheeting or panels in an attempt to add a superficial warmth or vitality to exteriors which were often raw, aggressive or humourless.

The postmodern period has seen a return of traditional values, context, humour, contradiction and complexity to architecture and facadism can rightly be regarded as an element within this process.

4

TOWNSCAPE PERSPECTIVES

Townscape is a subject which focuses on the visual character and qualities of the built environment. It is not so much concerned with the way individual buildings look in terms of architectural style, materials or age but rather with the essential character and quality of groups of buildings and the spaces defined by these buildings. It is concerned with effects such as novelty, surprise, beauty and orderliness. The theme running through the concept of townscape therefore relates to the three-dimensional visual impact of the built urban environment. According to Gordon Cullen, townscape is an art. In his book *The Concise Townscape* he refers to the art of the relationship between buildings and their surroundings. The process of townscape appraisal is not only concerned with the static relationship between buildings and spaces but also with the dynamic aspect of people and traffic movements. These are important elements which give an idea of scale and assist in the evaluation of the character, quality and ethos of townscape. Spaces only become alive when all townscape elements intermingle and are fused together.

Since the late 1960s in Britain the art of townscape has become increasingly pervasive in the world of urban town planning and design. It has also become a much more prominent element in architectural thinking than during the heyday of Modernism. While the art of townscape takes on a wider spatial perspective than architecture, the concepts of both are closely linked. Architecture is created within a context. Some architecture respects its context. Some architecture divorces itself completely from its surroundings.

During the Modern era the integration of new buildings with existing townscapes was not of fundamental concern. Architecture became a matter of designing in a completely new idiom. Buildings were to make bold independent statements unconstrained by past traditions. Topography, historic street patterns, existing scale and style, characteristic materials placed no inhibitions on the Modernists. The postmodern era has seen a move towards far greater contextualism in architecture and urban design. In Britain this process has received strong popular support and facadism has been an important concept within the process.

Facadism, in its various manifestations, raises a number of significant townscape issues. Facade preservation and replication may enable street scenes and vistas to be kept intact, although new roofscapes and replacement fabric behind the facades, unless hidden from view in the middle of a terrace, may significantly alter the appearance of the built form. Landmark buildings may be retained through internal gutting and remodelling. Some postmodern architecture may drastically change townscape views but retain contextual links with its surroundings through the use of surface symbols, motifs or decoration. Townscape grain may be significantly changed as a result of facadism.

Facadism may also affect people and traffic movements. For instance, in cases where two equally convenient routes through an urban area exist, and one route possesses far greater visual quality than the other, a significant proportion of route users may well be attracted to the higher quality route because it provides a much more satisfying visual experience. The route, therefore, may become a much-loved feature of a particular place and an essential ingredient of townscape character and ethos. The process of urban change and need for redevelopment may conceivably threaten these qualities. First, the removal of valued physical features may make particular routes less attractive, leading to changes in people movements along these and other routes. Facadism may be used to release the pressure for redevelopment while retaining the visual appeal of popular routes. However, the nature of remodelling behind preserved or replica facades may in itself alter people and vehicle movements, with a possible consequential change to townscape character, quality and ethos. A historic facade within a traditional shopping street may be preserved, for example, as the entrance to a modern shopping precinct developed on backland behind existing shops. As a result the traditional street scene is preserved but, potentially, the way that the local environment operates may be radically different. People may be drawn into the precinct rather than using the whole length of the original shopping frontage, with the effect that part of the frontage declines economically and premises are either put to other uses or fall into disrepair. The precinct is likely to have been built with its own car park and servicing facilities, which may have changed motorist movement patterns in the locality.

The preservation and replication of traditional facades is a course of action which relieves planners and architects of the responsibilities of agreeing appropriate and acceptable contemporary street frontages to new development. In the facadism of new postmodern architecture, superficial decorative features and motifs, drawn from past architectural traditions, have been used in an attempt to contextualise new buildings. The Terry Farrell Partnership's postmodern office block at Queen Street in the City of London, designed in 1986, provides a fitting illustration. Although a 'straightforward office development in terms of plan and accommodation, it is nevertheless important in the way in which it addresses the street in which it is situated. This is a well-mannered building, fitting neatly into its context despite borrowing from Manhattan and Viennese sources' (Glancey 1989).

The reawakened interest in the art of townscape as a prominent philosophy of British town planning is closely linked with the process of urban conservation. Certainly, conservation is an important process in the maintenance of high quality urban townscapes. Likewise, the art of townscape perception, which defines specific visual and spatial qualities, is a prerequisite of successful conservation. The philosophy of townscape and the process of urban conservation are not, however, the same thing. Townscape issues relate to the whole urban environment whereas urban conservation has a more specific focus on historic buildings and areas.

The debate concerning the townscape implications of facadism has generally concentrated on conservation matters, in particular the preservation of important street scenes and examples of formal urban design such as Georgian squares and terrace architecture. Here, however, the discussion takes a wider perspective and embraces townscape issues in general. Two of the greatest contributors to the philosophy of townscape have been Kevin Lynch and Gordon Cullen, and in considering the impact of facadism in relation to urban townscapes some of the basic concepts which underpin their ideas will be woven into the exploration of the subject.

Unlike arts such as music, the art of urban design can rarely use controlled and limited sequences. For different people and on different occasions sequences are reversed, interrupted or abandoned. Weather conditions and the time of day affect experiences. Not everything can be seen and heard at the same instant. There are hidden settings and views to be explored. An experience of one element of townscape is affected by other elements around. An experience is affected by sequences of events leading up to it and the memory of past experiences.

All mobile animals have a vital ability to structure and identify their environment. Visual sensations such as colour, shape, motion and light provide cues as well as senses such as smell, sound and touch. Disorientation within the urban environment brings a sense of deep anxiety and possibly terror. Such feelings reveal how closely people's ability to identify their environment is linked to their sense of balance and wellbeing. Therefore distinctly legible built environments offer security and also deeper, more intense, human experiences.

In his book *The Image of the City* (1960), Kevin Lynch used the results of an investigative survey of the role of environmental images on urban lives in three American cities – Boston, New Jersey and Los Angeles – to analyse visual elements in the city. On the basis of his studies, Lynch considered that the urban environment is broken down into five elements:

Paths – channels along which an observer moves, e.g. railways or streets, from which people observe the city while moving through it.

Edges – boundaries between two phases, e.g. edges of developments or walls.

Districts – areas of the city that people can mentally enter, which are recognisable through some common identifying character, e.g. similar facades, materials, colours.

Nodes – strategic points in an area which people enter into, e.g. an enclosed square.

Landmarks – a type of reference point such as a very prominent building.

The above elements do not exist in isolation. They regularly overlap and penetrate one another. For example, districts are defined by edges and structured with nodes. They may be sprinkled with landmarks and penetrated by paths. Lynch identifies other elements that are relevant to a detailed visual analysis of cities. Those relevant to our consideration of the impact of facadism on townscapes include spaces, frontages and individual buildings.

Lynch developed the hypothesis that an individual's knowledge of the city is a function of its 'imageability'. In a physical object, imageability is in Lynch's view a quality which gives it a high probability of evoking a strong image in any given observer. Shape, colour or arrangement facilitate the making of vividly identified, powerfully structured, highly useful mental images of the environment.[1] In developing his hypothesis, Lynch, while recognising that images are a compound of physical attributes and meanings, focuses on the role of form. A consideration of facadism in relation to Lynch's analysis of perception and townform raises some interesting points. Of particular significance is the role played by the facades of buildings as image elements in themselves and as part of a broader context.

Street frontages, for instance, may define a most important and highly popular path along which many people regularly move. The loss of such a path through redevelopment may seriously affect orientation within their particular urban area. In the event of redevelopment pressure being irresistible, facadism offers the opportunity of preserving the street frontage with redevelopment taking place behind, in which case the path may be kept intact. Of course, the construction of new buildings along the same building line as the older buildings would preserve the path. However redevelopment would affect people's observations from the path and result in different cues, messages and meanings being conveyed, with a possible disorienting effect. By contrast, a decision not to preserve existing facades may result in the formation of a new, more imageable urban path which increases people's orientation.

The practice of retaining or replicating existing facades may help to protect the identifying character of a particular district. In the New Town area of Edinburgh the late eighteenth- and early nineteenth-century developments possess a remarkable unity of design. It is a district wherein one is aware of a common identifying character. The ordered architectural rhythms and the characteristic colour and texture of the stone displayed on the facades of the buildings play an important part. Already, facadism has played a role in preserving the identity of the district, particularly in George Street which formed the main axis of the earliest New Town development, begun in 1767.

Again, in Bristol, the Portland Square conservation area comprises a townscape district within which the imageability of the locality rests on the planned uniformity of the Georgian development. The identity of this district has been enhanced as a result of numerous developments throughout the 1970s, many of which have been cases of facadism. Some facades have been preserved, others have been replicated. On the edge of the district, adjacent to the M32 motorway (the main northern route into and out of the city), new buildings have been erected evoking the traditional style of buildings within the district.

There are two strategic points or nodes within the Portland Square conservation area, one being Portland Square itself, and the other, Brunswick Square. These squares are fed by numerous routes through the area and provide travellers along the routes who enter these nodes with strong mental images of their journey. Brunswick Square was almost lost during the 1960s when the City council was promoting the demolition of the southern side of the square in preparation for the continuation of modern office development, as part of the outward expansion of the City's central business area. However, the proposals never came to fruition and sufficient of the square has survived for it to remain an important strategic point. In the early 1980s the facade along the southern frontage of the square was retained as part of an office redevelopment scheme.

Landmark buildings present powerful visual images and are valuable identifying townscape elements. At the western end of George Street, Edinburgh, the vista is closed by the dominant landmark of a former church building (see Plate Section, viii, xi and xxii), which has been completely remodelled inside to form a public records office. In Washington DC, the seven-storey 1901 Bond Building was a prominent corner-site landmark building. During the 1980s the two beaux-arts facades were preserved and the Sigal/Zuckerman Co. constructed a twelve-storey building behind the facades, with the five storeys that project above the preserved facades being set back twelve feet. Also in Washington DC and on a similar theme, a six-storey facade, retained at a corner site in Pennsylvania Avenue NW as part of the frontage of a modern office development – which rises a further three storeys above the preserved fabric – provides a strong visual townscape cue (see Plate Section, xii).

Occasionally, freestanding facades or parts of facades have been preserved as sculptural landmarks. During the early 1970s, in Independent Square, Philadelphia, Penn Mutual Life Insurance Company dismantled, at a cost of $130,000, a four-storey marbled facade dating from 1902 and bearing Egyptian revival-style detailing. After a period of three years the facade was rebuilt as a freestanding sculptural wall over which the new 21-storey Penn Mutual building was cantilevered. In Portland, Oregon, the facade of a fire-damaged building dating from 1892 was preserved to screen a car park. The stone and brick facade with arched windows and terracotta ornament was considered to be important to the overall texture of the street.

At Marples Wharf, Bath, a series of stone arches, which appear to have been part of an elevation of a former riverside industrial or warehouse building, have been preserved as a feature which allows views through the arches to the townscape beyond. The arches are an attractive and interesting townscape feature in themselves and provide a memorable landmark in the locality. They also convey information. The arches have been preserved as a piece of sculpture to adorn the setting of the adjoining building, which has been converted to offices (see Plate Section, xviii). This tends to inform the observer that the former industrial character of the locality has now been transformed into the domain of the white-collar worker where the provision of such adornments has traditionally been more common than within the industrial landscape. The case at Marples Wharf is similar in nature to the granite arch which stands alone as a monument to the past at Albert Dock, Liverpool. The arch was once the entrance to a former tram shed, but is now a landmark and a symbol of the former use of the dock, which has been rejuvenated as a mixed cultural and commercial enterprise.

The use of parts of former buildings as urban sculpture, symbols and landmarks could be regarded as a somewhat controversial subject. However, such features provide reminders of the past, create interest and possibly beauty and give townscapes greater identity. For many people, if not all, such features will facilitate a more comprehensive perception of the built environment.

The above reflections indicate that in some circumstances facadism may be important to imageability in terms of physical urban form. Certain facades or groups of facades are likely to facilitate vivid, powerfully structured, invaluable mental images of the built environment. The practice of preserving or replicating existing facades to screen new development or of designing buildings which outwardly evoke the style of surrounding traditional buildings is symptomatic of a continuing reaction, at least in the UK, to the brutal experiences of the Modern era. In the USA, cases of facadism may reflect the relative shortage of historic townscape. One of the reasons why Modernist townscapes have become so disliked is the fact that they are monotonous. One of the qualities of Modern architecture was its functional simplicity and clarity of design. Imageability, however, is derived from a continuing interest in the physical form brought about by a degree of complexity or ambiguity and thus a certain lack of clarity. Such complexity or ambiguity may often be more prevalent in townscapes which have evolved over the centuries, displaying a diversity of design styles, rather than in the transplanted townscapes which reflected the uniform Modernist dogma.

Having recognised this point, it may also be argued that on occasions a slavish preservation of existing frontages and attempts to evoke the established urban form may deny opportunities for achieving a higher degree of imageability within particular townscapes. Two buildings with the continuing power to shock and stimulate are the Lloyds building in London and the Pompidou Centre in Paris. These buildings challenge the established order of adjacent townscape and inject tremendous ambiguity and complexity into the urban scene. The Pompidou Centre is a 'vast exercise in Bowellism' where the building weaves its insides outside (Glancey 1989). Both the Lloyds building and the Pompidou Centre are clearly controversial buildings because of their dramatic townscape impact. An alternative option in each case would have been to provide more modest contemporary interiors within a traditionally designed shell. However, the townscapes of London and Paris are vast enough to accommodate such extravagant experiments. The fact that the Pompidou Centre has remained as popular as the nearby Eiffel Tower demonstrates that the building commands a continuing interest among vast numbers of people, which is partly a product of the complexity and ambiguity of the resultant townscape.

Such qualities are important components of imageability.[2] Complex and ambiguous townscape elements need not be necessarily as dramatic as the Pompidou Centre, although they usually bring to townscape the quality of surprise. For example, in the centre of the Cornish coastal town of Penzance an extraordinary early nineteenth-century building known as the Egyptian House brings a touch of colourful fantasy to an otherwise plain townscape. At Broad Street, near the commercial heart of the city of Bristol, a delightful tiled facade to the former Edward Everard printing works has been retained as the entrance to a modern office complex. The original building was designed by Henry Williams, with the decorative scheme on the facade by W. J. Neatby, chief designer for Doulton. The intricate tiling work, in white faience and highly coloured Doulton Carrara ware (named because of its resemblance to Carrara marble) was by Walter Yoxall, a local craftsman. Featured in the design are Gutenberg, inventor of the first method of printing, and William Morris, the artist, theorist and leader of the renaissance of fine printing during the Victorian era. The facade is a masterpiece of English Art Nouveau and the complex arrangement of shapes and colours creates a stunning effect of tremendous townscape interest. Also the facade is set back from adjoining developments, which provides an additional element of surprise to the observer. All these factors combine to form extremely useful mental images and this example underlines the important role that facadism can play within the art of townscape (see Plate Section, xv).

The important townscape concept of serial vision has been described as an observer's experience of perceiving townscape as a series of revelations when moving through an urban area, involving both existing and emerging views (Cullen 1971). These views may be natural or carefully planned, close range or long distance across a townscape. In many situations a facade or groups of facades provide highly significant elements in attractive urban views. For example, approaching Bath's Royal Crescent along Box Street from the east, the slowly unfolding view of the great curving facade which is revealed to the observer provides a highly dramatic townscape experience. The position of facades and their relationship with their neighbours, some being set back or forward, others being seen from oblique angles, others face on, provide a rich diversity of views within historic townscapes. Facade preservation can be an important means of ensuring that such diversity is retained, because it means that developers must work within the constraints of existing street frontages, whereas if historic buildings are demolished in their entirety it is more likely that the characteristic and subtle juxtapositions of buildings will be lost.

Longer range views are also significant in the perceptive experiences of observers on their journey through townscapes. Bath's Royal Crescent, because of its elevated position and open aspect in front, provides outstanding longer-range views from various parts of the city. Such important views raise a relevant issue in our exploration of the townscape significance of facadism. In the event of the fabric behind the facade of the Royal Crescent requiring replacement, it would be essential to the townscape character of Bath that the facade should be preserved or at least rebuilt. The same could be said of numerous other examples of terrace architecture throughout the UK, where buildings were erected primarily as pieces of townscape scenery. A particularly impressive facade in the context of urban views has been preserved in front of a contemporary office development at St George's Road, Bristol. Here, the facade of Brunel House, because of the openness of the townscape in front of the building, is clearly visible from numerous vantage points and the monumental classical style edifice with its long portico of powerful fluted Ionic columns commands the attention of onlookers (see Plate Section, vi).

Less pleasing are the views available of Brunel House from higher ground, which reveal the awkward roofscape of the new building constructed behind the facade. The present roofscape lacks the intricacy and visual interest of that which it replaced (see Plate Section, xix) and tends to over-dominate the scene from certain angles. This is particularly disappointing to the observer who approaches Brunel House, descending Brandon Steep immediately to the west and an interesting townscape route from the nearby Brandon Hill, one of central Bristol's most popular public open spaces. The Brunel House scheme provides a convenient illustration of the cost and benefits involved, in townscape terms, in adopting facadism as an approach to urban remodelling.

When particular urban spaces allow individuals to identify with that space – perhaps through evoking a sense of drama or tranquillity or because of aesthetic qualities – the important townscape notion of a sense of place is bestowed upon the space. The experience of entering an urban square is often one of drama; this may be most pronounced in cases where there is a strong feeling of formality and enclosure. Within the squares, the whole composition is more important than the individual buildings. Three elements normally come together to create the whole: individual buildings; the grouping of these buildings; and the central open space. In the case of British Georgian urban design the harmonisation of style, materials and landscaping was an important concept. Bedford Square in London provides an example of an elegant square of Georgian houses. Here, anything other than preserving the harmony of the square would destroy the composition. In cases such as this, where buildings require remodelling either for reasons of economic necessity or structural decay, the concept of facadism offers a valuable option in terms of retaining townscape quality and a sense of place. This approach was used on a number of occasions at Portland Square, Bristol, where, to a considerable extent, facade preservation and replication as part of office remodelling schemes have retained the special sense of place within the formal Georgian square. Facadism may be more difficult to justify in places such as Trafalgar Square in London, which comprises buildings of a diverse range of styles and sizes. This is a diffuse composition where there are greater opportunities of introducing new styles.

The impact of vehicles on townscape is immense, particularly in terms of providing parking within town centres. Multi-storey car parks, in particular, can have a devastating visual effect on the urban scene. In Kansas City, a seven-level car park to accommodate 2,200 vehicles in a historic area of the central business district has been partially masked by the retention of the nine-storey facades of two buildings. In Portland, Oregon, a stone and brick facade with arched windows and terracotta ornament was salvaged from a fire-damaged building dating from 1892 and used to hide a parking lot.

Physical attributes such as architectural detail and prominence contribute to but do not guarantee imageability and a place in an observer's mental model. If a sense of place is to be felt amid townscape, senses other than the visual and meanings other than the concrete may have to be evoked. Thus townscape may be enjoyed for its own sake at a preliminary and temporary stage of sensation, but if physical form is to become an indelible imprint for the observer and a real quality of imageability is to be achieved, it would seem that the visual experience must be supplemented by some level of meaning experience. 'It is the attribute of significance which gives meaning to environmental happening' (Ittelson 1960). This is where buildings which have special meanings to the community or act as symbols, both at a national and local level, are so important in generating a sense of place and strong qualities of imageability. The Houses of Parliament, the Royal Albert Hall, the Louvre, the White House and La Scala may be attractive architecturally, but it is the cultural significance of the buildings which imprints them in people's mental models. At a more local level, all sorts of meanings and associations are attached to areas of townscape. For instance, particular buildings may be of great significance in employment terms to large numbers of a local community. Other buildings may be relevant as places of public assembly.

Some buildings may have specific historical associations such as being the birthplace of a famous person or the location of a significant event. At the level of the individual, certain townscapes may generate a particular sense of place because they have provided the setting for an especially meaningful experience. In considering the role of facadism in the process of managing townscapes, it is particularly relevant to recognise the importance of historic facades as components of an image. Within many townscapes the facades are the only parts of the buildings that people ever experience and so possess a special importance in this respect. Historic townscapes have provided the backcloth for vast numbers of meaningful experiences, which have been passed down from generation to generation, and so the value in terms of significance of such townscapes has been accumulated through the passage of time. The destruction of townscapes, therefore, to make way for redevelopment may erase highly valuable meanings and associations attached by people to townscapes with a consequent loss of identity with urban areas. In addition, the architectural detail and aesthetic qualities of historic townscapes are often irreplaceable either for reasons of cost or lack of expertise, and so demolition may destroy both the visual and meaning experience. It is possible to retain or enhance a sense of place, or preserve meanings and associations through facadism. In cases where the process of urban change is irresistible, the preservation of architecturally fine facades may be sufficient to enable valuable visual experiences to continue.

Also, by keeping street scenes intact the necessary cues and reminders to facilitate meaning experiences other than the concrete are being provided. For example, the facade of Brunel House in Bristol is not only of immense architectural interest but has important historical associations with the engineer Isambard Kingdom Brunel and the Great Western Railway. The original building was constructed between 1837 and 1839 as the principal terminus hotel of the Great Western Railway. Therefore the building's dominant physical presence is supplemented by its associations with one of Britain's most brilliant pioneering engineers of the past. The preservation of the facade may also provide an intricate web of more specific associations, such as providing a link for observers whose relatives may have stayed at the Royal Western Hotel or worked for the railway company or under Brunel himself. Such associations give depth to the mental images of townscapes held by observers.

The facades of buildings are powerful elements in displaying the fabric of settlements, including colour, texture, grain, age, scale, style, character, history and uniqueness. In continuous street frontages the facades of buildings are the fundamental element of buildings which display the characteristic colours and textures of the materials and create local identity. Historic facades bring the pattern of age, which is irreplaceable. This is an important consideration when it is proposed to demolish an existing historic facade and replace it with a replica newly built version. Although the latter may be helpful to a developer in not having the constraints of working around a supported facade, the fact that historic facades reveal the signs of age and weathering is often part of their particular appeal. This and other pitfalls of creating replicas are referred to by Anthony Tugnutt and Mark Robertson in their book entitled *Making Townscape* (1987). The problem is not being able to replicate fully the appearance of the former facade in newly built replicas with present-day construction methods and the need to comply with building regulations. The original facade is likely to have been of load-bearing construction, whereas the replicas will be thin skins of stone or brick in front of a contemporary building. Another problem referred to is that cost considerations may result in the use of artificial rather than natural stone. Windows may often be the subject of insensitive changes such as the depth, size of window opening or dimensions of component parts. The cumulative effect of these matters may well result in a facade which is only a pale image of the original.

Facadism may be used as a mechanism for ensuring that the scale of new development is in context with the surrounding townscape. For example, where town planning decision-makers require that existing facades are retained or replicated and that new development behind the facade should respect its scale, then developers are constrained to working within an established framework. Where, however, facadism takes place without such constraints and the new fabric breaks out of the traditional shell, there may be some bizarre consequences such as the following example in Cambridge, Massachusetts:

> a bay was removed from a side of the 1932 granite East Cambridge Savings Bank building and used as the front facade for a new addition, a curving steel and glass corridor that flowed from the original building as if it had sprung out, pushing the granite facade with it.[3]

Also in the USA, there have been cases where facades of historic buildings have been retained at the base of new multi-storey buildings. Roy Solfisberg, the architect in charge of a project for Holabird and Root, the firm that in 1912 originally designed a building in Chicago where the granite and terracotta facade of the first four floors has now been preserved in front of a 37-storey building racing skyward, takes the view that the tower block is little noticed as part of the streetscape as the majority of people are aware only of the lower 40 feet of a building.[4] An observation of a crowded urban street will confirm that most people do not habitually look up at the upper storeys of buildings. However, the impact of new fabric considerably higher than a preserved facade will depend on the number of similar-height buildings within the immediate townscape. A constant sense of enclosure created by many tall buildings of similar height is less likely to cause people to look up rather than a sudden discordant element. The reason given by Roy Solfisberg for the retention of the facade was that it 'maintains a quality level that is very difficult to replicate in modern buildings'.[5]

In historic British and European towns there is often an established hierarchy of size and shape which so eloquently expresses the individual and collective functions of buildings. The hierarchy was not respected by the high-rise buildings of International Modernism. In the historic town or city the satisfying skylines were created by the relatively uniform size and materials of small buildings, which gave deference to the larger institutions such as castle, church and town hall. These rose above their neighbours, reflecting their importance within the prevailing ethos. If urban change takes place within this ethos the visual character of townscape is retained. It is a somewhat bizarre concept to preserve facades of historic buildings, which were constructed within the framework of an established ethos, and then to dwarf the facades with buildings that clearly have nothing to do with such tradition.

This brings us to the question of authenticity and setting within townscape. Dr G. J. Ashworth considers that the 'question of authenticity is raised as much by context as by historical detail'.[6] He refers to the shortage, by European standards, of available surviving buildings of note in the North American city of Vancouver, which frequently results in great incongruity between preserved historic fabric and new development:

> A mid-nineteenth-century Anglican church, one of the oldest buildings in Vancouver, is carefully preserved in a context of surrounding high-rise modern buildings. In fact the planning instrument of development rights transfer, where the conservation of the church may be financed by the sale of its site development potential to its neighbours, will exacerbate the contrast.

> The authenticity of the resulting townscape can be questioned and evokes little of the Victorian city.
>
> (Ashworth 1991)

This issue of authenticity and setting is evident in the case of a small, elaborately detailed, Classical-style facade which has been preserved as the centrepiece of a small 1970s office building at Redcross Street in Bristol (see Plate 8). The new fabric of the office building respects the scale and, to some extent, the proportions of the Classical facade. However, the protruding smooth cement-rendered cheeks of the modern fabric appear sadly inappropriate compared with the preserved facade. The subservient role of the historic fabric in this case is further emphasised by the fact that the preserved facade is set back between the neighbouring wings of the new office building. Towering menacingly above the small office building is the 1970s high-rise development. The lack of other historic fabric of the quality and style of the preserved facade in the immediate vicinity means that it is bereft of an appropriate context and setting. The office building, which wraps itself around the facade, appears as a clumsy attempt to contextualise the remaining lonely historic element. The window detailing, materials and horizontal emphasis in the courses between the panels are brought down from the tower block to form a three-storey version, with the later design feature reflecting the storey levels on the historic facade. The facade itself is an interesting townscape element and people using Redcross Street as a route through a rather barren and austere urban scene may welcome the visual relief and reminder of an artistically finer past when buildings were constructed to a more human scale. Clearly there is no historical authenticity or setting to this piece of facadism. The feeling conveyed is that the architects have felt obliged to save the historic facade, but have not been able to solve successfully the problem of accommodating it in the new development.

The preserved six-storey facade at Pennsylvania Avenue NW, Washington DC (see Plate Section, xii) provides a further example of historic fabric being retained in a setting which provides little contextual relationship. While the facade is a strong visual element in its own right, there is an incongruity in the way it has been overwhelmed by the stepped structure behind.

In some cases, facadism represents a truly inclusive approach to remodelling townscape where historic elements are retained, replicated or evoked within contemporary townscapes either to add interest and decoration or act as symbols, monuments and reminders of the great architectural traditions. If this view is taken the argument emerges that facadism in the form of preserving facades or creating replicas has a similar townscape impact to Berthold Lubetkin's entrance canopy to his flats at Highpoint II, which features exact copies of details from one of the buildings standing on the Acropolis in Athens (see Plate xiii in the Plate Section). The inclusion of the preserved facade at Redcross Street in Bristol would have featured more meaningfully in the local townscape had it been given a more dominant role within the new development. Whereas Lubetkin's caryatids are classical features woven meaningfully into an other-

Plate 8 Redcross Street, Bristol.

wise modern development, the preserved facade at Redcross Street has been aimlessly sandwiched into an annexe to the main development. The result, in townscape terms, neither captures the apparent essence of Modernism nor contextualises successfully with the preserved fabric.

Where facadism is part of an inclusive approach to managing townscapes it helps to reflect the complex and contradictory nature of urban areas. Many of the urban townscapes of Britain, Europe and the USA have become mosaics, displaying the relics of historical tradition standing alongside the products of a now broken optimism of the Modern era and the subsequent inclusive and contextual experiments of post-modernism. The concept of facadism is based on demands within contemporary societies to include, on the one hand, the visual quality, interest, historical associations and nostalgic reassurance provided by traditional images and, on the other hand, the advantages, convenience and comfort of present-day accommodation and technology.

The Danish town of Ribe can be considered to be a small conserved Western European gem containing the largest concentration of listed buildings outside Copenhagen. Ribe's location on the Baltic–North Sea trade routes endowed the town with great commercial prosperity during the sixteenth and seventeenth centuries. A period of stagnation following the shifting of trade routes subsequently stifled pressures for development. The imposition of controls over change and development prior to the onset of Modernism meant that the completeness of its historic fabric remained

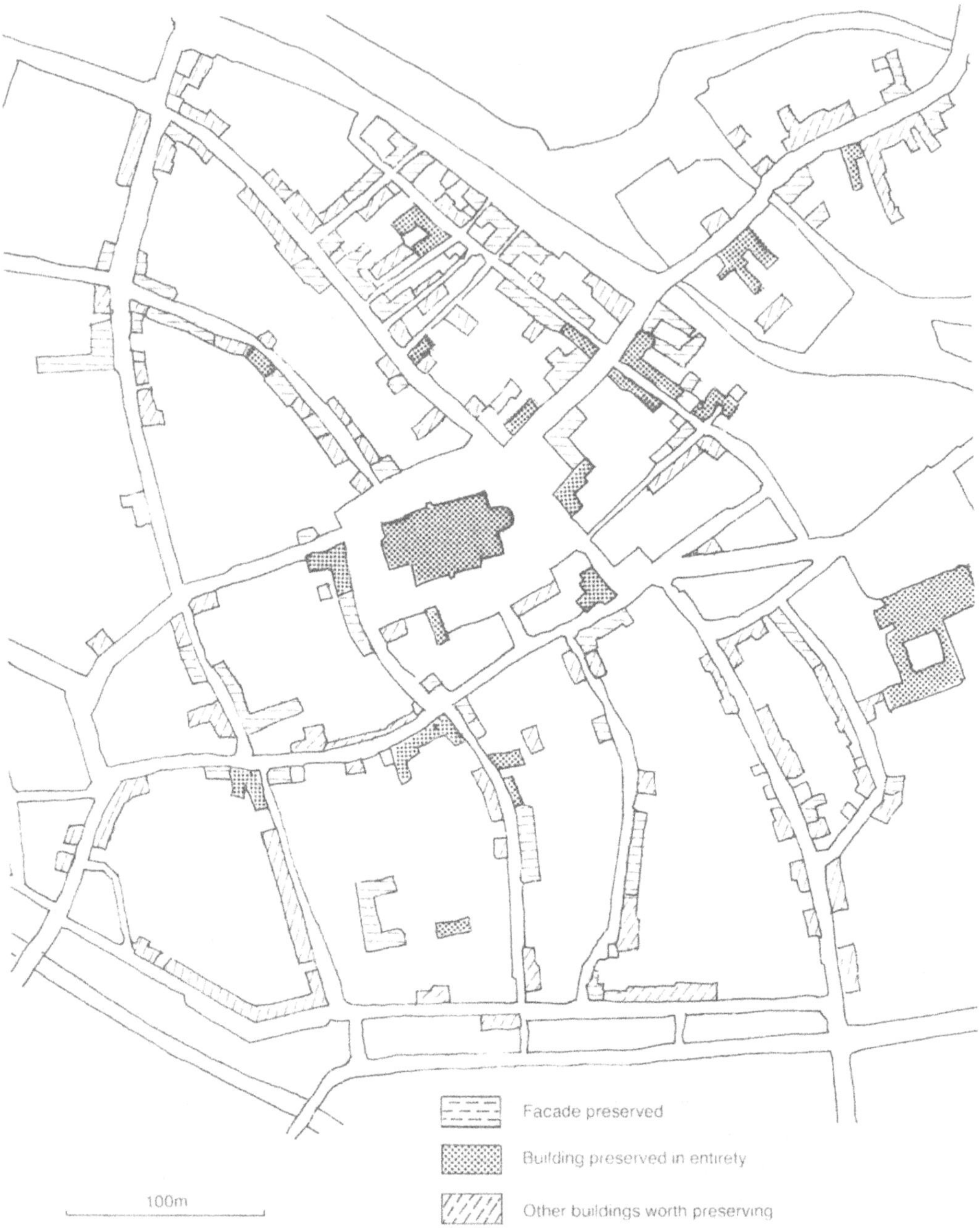

Plate 9 Centre of Ribe, Denmark, based on original in G. J. Ashworth (1991).

relatively intact. It is interesting that facade preservation has been used extensively in Ribe to preserve the visual quality of its townscape (see Plate 9). Clearly, the decision-makers responsible for managing the townscape in Ribe consider facadism an acceptable approach and, judging by the town's appeal to visitors – it attracts around a million visitors each year – this is a most successful policy.

The concept of facadism, therefore, is closely linked with the processes of managing historic townscapes, urban conservation and heritage planning. There are two aspects to be considered here. First, there is the relationship of facadism to the management of mono-functional heritage towns and cities that have survived intact without the intrusion of unsympathetic forms of development, such as historic gem towns and cities where the whole ethos of the places, their economies and identity, are dominated by the historic resources. In Europe, classic historic gems would include as well as Ribe, Willemstad in the Netherlands, Neuf-Brisach and Aigues-Mortes in France and Dubrovnik in the former Yugoslavia. Even the new world contains a few, such as Louisbourg in Canada, and St Augustine and Nantucket in the USA (Ashworth 1991).

Second, there is the role of facadism within multi-functional towns and cities, where Modern and postmodern development has changed the nature of places but historic townscape elements are still important to the identity of places and to their economies. In multi-functional towns and cities, where the historicity of the place has been fractured by Modern and postmodern forms of development, it is more likely that incongruous forms of facadism will occur. This is because the diversity of the social, economic and functional base of the town or city is not solely based on its heritage and resources. Consequently, it is more likely that, on occasions, pressures for change and development will override conservation principles and result in development that is not truly authentic to its setting. In such cases – for example, that referred to above (pp. 66–7) relating to Redcross Street, Bristol – facadism is not using heritage resources in the true spirit of conservation but as a means to embellishing contemporary townscape.

The pervasive spirit of urbanism which, throughout Britain and Europe, has swept away much of the dogma of the Modern era has restored the townscape importance of the facade. Modern architecture possessed an anti-urban quality and buildings were designed without recognisable facades. The anonymous functional structures screamed for attention in isolation from the harmonious fabric of traditional town and city streets. Within the re-establishment of urban values, architectural and urban design thinking now respects the prevailing traditional ethos. The art of the well-mannered facade is once more a relevant design issue, as it had been during the eighteenth and nineteenth centuries when buildings paid respect to one another. Interestingly, in the context of this book, it was not unusual for facades to hide a wealth of interior plans. In London, for instance, we would be just as likely to find a small, elaborately embellished palace complete with its own ballroom behind a plain brick house facade as to discover an ordinary merchant's dwelling (Glancey 1989). When seen in this light the concept of facadism sits comfortably within the prevailing spirit of urbanism.

Plate viii George Street, Edinburgh.

Plate ix Ca'D'Oro Building, Union Street/Gordon Street, Glasgow.

Plate x Photograph of Ca'D'Oro Building, Glasgow.

Plate xi Nos 4 and 5 St Andrew Square, Edinburgh during redevelopment in 1992.

Plate xii Pennsylvania Avenue NW, Washington DC, USA. Photograph by Professor Paul Knox, Director of the Center for Urban and Regional Studies, Virginia Polytechnic Institute and State University, USA and included with his kind permission.

Plate xiii Lubetkin's entrance to Highpoint II.

Plate xiv The Arnolfini Building, Bristol City Docks.

Plate xv An elaborate Art Nouveau facade in Bristol.

5

SOCIAL AND ECONOMIC PERSPECTIVES

Much of the debate surrounding facadism has in the past focused on architectural and townscape issues and its implications in terms of the urban conservation process. These subject areas are, of course, embroiled in the urban system in its widest sense and, consequently, raise various social and economic considerations. Every development decision, even concerning the most seemingly insignificant matters, involves a balance between a range of social and economic costs and benefits. For instance, in the simplest form of facadism, where a person refaces the front elevation of a house with artificial stone cladding, the financial costs of obtaining and fixing the new material are weighed against the pleasure derived by the householder at the new appearance of the dwelling. There may be marginal savings through improved insulation. Certain members of the community may feel that the alterations to the house are distasteful or out of keeping with the locality and in effect may suffer social costs. The action by one person may stimulate others to do the same, thereby compounding these social costs. In areas of architectural or historical interest such proposals may be contrary to the conservation policy objectives of planning authorities, and may create costs in terms of officials spending work time in negotiations or investigations. In the case of larger development projects the costs and benefits are highly complex and extensive. Other writers have dealt with the techniques of cost–benefit analysis and it is not my intention in this book to quantify the costs and benefits involved in cases of facadism. Neither has my research for this book included detailed analysis of popular attitudes towards facadism. My approach has been, on the basis of my experience within the town and country planning system, to introduce a range of ideas which should be part of a balanced consideration of the subject.

SOCIAL PERSPECTIVES

All major development work causes disruption of varying degrees to communities and, consequently, social costs. Where existing historic facades are preserved *in situ*, initial social costs tend to be high. As it is rarely possible to provide the necessary support from within the development site, the elaborate scaffolding structures required to support safely fragile skins of stone or brick frequently block pavements or intrude into roads, causing disruption to both pedestrians and motorists. Pedestrians, therefore, may be forced to cross busy roads to avoid obstructions and traffic may have to be diverted. The inconvenience can be costly both in terms of frustration and stress caused to the public, with the health implications involved, and in delays to traffic movements, particularly if the preservation of facades is a long-term matter. Demolition and redevelopment may be far less disruptive in this respect, because the activities can probably be contained more within the development site and, because there are fewer constraints, such as working around a preserved facade and linking it to the new structure, the project is likely to be completed more quickly.

It is inevitable that, from time to time, the urban scene will be disrupted by building works. This is one of the ongoing costs of environmental damage to buildings and of the pressures for urban change. On occasions in recent times, development sites in some UK cities have been enclosed behind hoardings displaying murals or artists' impressions of the development taking place, supposedly to lessen the visual impact of the construction works. During the 1992–93 football season, Arsenal FC took the trouble to have the North Bank at their Highbury ground redeveloped behind a crowd-scene mural. It is assumed that the club considered its team would perform less effectively without the vision of supporters baying behind the goal at the North Bank end of the ground, or, perhaps, more altruistically, it was felt that spectators would appreciate the artist's attempt to create the feeling that the North Bank is still a throbbing mass of humanity rather than a building site. I have not attempted a study of the reasons behind the practice of erecting these screens, but it would be an interesting subject area. Perhaps it is indicative of a concern of developers over the visual costs to communities of construction schemes. Alternatively, it may be part of the same culture as that which leads to facadism, in that there is a self-consciousness or fear among those responsible for contemporary buildings, which leads to a wish to hide them – either during construction behind murals, or permanently behind historic or facsimile facades.

In terms of the visual costs involved in new development schemes, facadism is little different to restoration works, reconstruction or new building. Preserving historic facades may involve marginally longer projects because of the difficulties of working around a supported, delicate skin of stone or brick and then attaching a new structure to it, but the public is unlikely to have to endure the sight of construction works for vastly longer periods. Another aspect to be considered here is that some people, rather than finding development an unsightly and unwelcome disruption, regard it as a fascinating activity or an encouraging sign of progress. Also, some of the elaborate scaffolding structures created to support facades are almost technological art forms in their own right (see Plate 10) and can be interesting to observe while they are in place.

In cases of facadism, the preservation or replication of certain historic facades, because of their architectural quality or townscape importance, may represent a community gain which mitigates the initial social costs endured by the community. Several typical social costs and benefits produced as a result of facade preservation are

Plate 10 Typically elaborate scaffolding structure, Redcliffe Street, Bristol, 1988.

illustrated in the case of York Street, Bedminster, in Bristol, where the facades and a number of internal cross-walls of a prominent terrace of former Georgian town houses have been supported by scaffolding for well over 12 months. This terrace has been derelict for many years, unprotected from the elements, and it is doubtful that any fabric other than the facades will be restorable. One of the costs to onlookers is the visual bleakness about the site and the atmosphere of dereliction and recession overhanging the locality. Also, the pavement in front of the buildings cannot be used and the highway has, in effect, been narrowed by the scaffolding structure, restricting traffic flows and thereby contributing to delays, particularly during peak hours. York Road runs alongside part of Bristol's network of waterways and was once a fashionable place to live. It retains many dwellings along its length, but is now regarded as a location at the lower end of the housing market. The traditional facades of York Road overlooking the river, although not outstanding architecturally, are important contributors to the local scene. In recent years some unremarkable infill development has taken place along York Road between existing buildings which, although respecting the scale of the terraces, neither replicates previous facades nor makes an outright contemporary statement. The result is a depressing mediocrity which tends to justify the preservation of existing facades despite the current costs.

In *The Heritage Industry* (1987), Robert Hewison expresses the view that Britain has developed a heritage culture and one of the consequences is the reinforcement of social values which, in his opinion, the democratic advances of the twentieth century were eradicating. We should consider whether or not facadism is part of this culture and the extent to which it could be regarded as a socially regressive practice. The urban conservation process is concerned with preserving townscape images of the past and the historical associations attached to buildings and places. Facadism is an important part of this process. Many of the townscapes now treasured by the present generation were created out of societies divided by class, privilege and social injustice. Consider, for example, the richly ornamented historic commercial buildings in the towns and cities of the UK, which were lavishly funded by industries and trades while the labour force lived in abject poverty and human rights abuses were appalling. An extreme view is that the buildings created out of such socially divisive conditions convey undesirable images and messages within the built environment and should be replaced by buildings representing a more democratic system. Adopting such a policy would remove so many treasured historic buildings as to be widely unacceptable, at least in Western cultures. Reuse of such buildings offers an opportunity to find more socially acceptable uses for them. It could be argued that facadism potentially enables an even more complete redemption by allowing the retention of what are often the most architecturally valuable parts of buildings – namely the facades, which can greatly contribute to the quality of life in our urban areas – and the incorporation of these in what is essentially a new creation representative of, it is hoped, a more democratic society. Facades which may have had negative associations are then possibly transformed in terms of their symbolism.

It cannot be claimed that the UK is now a truly democratic society. There are still far too many social injustices where people are manipulated by powerful interests, society is not classless and too many minority interests are either not represented, or they are ignored. However, British society has, in many ways, become more equitable since the 1930s, and it is important that new innovative architecture is given an opportunity within major British city and town centres to inspire, entertain and impress present and future generations even alongside the great architectural creations, and within the treasured townscapes, of the past. While historic environments do not inevitably inspire a wish in people to return to the social practices of previous eras, if contemporary architecture and urban design are constantly denied in favour of preserving elements of the past, there is a danger that society will feel that the present has little of architectural or townscape worth to offer and that past social values should be reasserted to enable the creation of good quality urban environments.

Such an argument would appear to carry greater weight in times when societies are sufficiently lacking in confidence in their ability to deal adequately with the present so that the past becomes excessively glorified and an unbalanced influence on current events. This is one of the concerns about the UK at present and the impact of the heritage industry in distorting the real images of the past (Hewison 1987). It is unfortunate, in the case of the UK, that the architecture and town planning of the post-war era – the products of an age regarded as relatively socially progressive – caused such widespread disillusionment within society and a lack of faith in contemporary development to provide an acceptable townscape solution. Some, although not all, cases of facadism reflect this attitude. The desire to retain historic facades of undoubted architectural quality, townscape impact or historical significance is understandable and socially justifiable given their intrinsic value and contribution to the quality of people's lives. However, where decision-makers lack the will and courage to occasionally allow the replacement of historic facades with high-quality contemporary designs, it remains difficult to re-establish the feeling that the current age has a valuable contribution to make to the built environment. There are situations where, through sensitive redevelopment, townscapes may be better enhanced than by preserving traditional facades. For example, many historic facades are by no means unique in architectural or historic terms and are often badly disfigured by weathering and decay. In some cases, they may be ill-at-ease within their surroundings or even represent incongrous townscape elements because of their scale or style. It is, I believe, socially important to achieve balanced townscapes which represent the highest achievements and aspirations of every age. The built environment should enable men and women to look back to their roots, feel part of the present and, if they wish, imagine the future. There is a place for facadism, but by constantly resorting to it to solve the question of the street frontage are decision-makers not denying their responsibilities to present and future generations?

The distinction between preservation and conservation of the built environment is important to this discussion of social perpectives. Conservation is a dynamic process which allows places to adapt to contemporary social trends, although the past remains as an enclosure or setting for new ways of life. Vital links with the past are retained to meet people's psychological needs. Preservation, on the other hand, captures the past and brings it forward into the present unchanged. On occasions, urban conservation will involve the complete preservation of important fabric, but more often some adaptation of varying degrees, including drastic action such as facadism. However, it is when attitudes towards the built environment are excessively preservationist that the dangers of regressive social values from the past being reasserted will be greatest. Eversley, for instance, points to the fact that development choices often lie between preserving an asset for the benefit of a few and improving the living standards of many: 'Small shops stock specialities for a tiny wealthy minority or even tinier intelligentsia. Supermarkets may be "brash" . . . but they reduce the cost of living for millions.'[1]

There is an argument flowing out of this that social injustice is imposed by powerful minorities who regard old buildings, existing townscapes and traditional uses as intrinsically good. In the light of such a view, a preservationist approach to conservation is elitist and undemocratic and an activity which can, in certain circumstances, prevent the improvement of the built environment for the benefit of less fortunate sections of communities. For example, the retention of old, obsolete buildings of inefficient design and little economic use may prevent the provision of improved residential or employment-generating development.

Facadism, in terms of urban conservation practice, is a drastic concept, but, it could be argued, is less socially regressive than many other measures, given the flexibility it allows in adapting historic areas behind a preserved street scene. To take Eversley's criticism, for instance, that the conservation process restricts the benefits of cheaper shopping in supermarkets from reaching large sections of communities through protecting small old shops, it has been possible in some historic areas to carve out the space behind a series of traditional street facades within which a supermarket or 'magnet' store has been located, thus bringing the economic and other benefits to large numbers of the local population.

Many towns and cities in the British Isles have examples of this kind of development. I have, for example, recently encountered town centre remodelling schemes at Bath, Cheltenham, Exeter and Barnstaple,[2] where the frantic modern shopping experience takes place, at least in part, behind the reassuring packaging of historic facades.

The social costs of drastic change should not be overlooked. Where historic fabric is swept away, so too may be the communities that once inhabited a locality, or at least vital services to those communities. To remain with the example of shopping, small old shops sometimes serve a vital need for those who are unable to travel to supermarkets, or for people who by choice use smaller shops. If quality of life is to be fully protected, choice must be offered. Writing in 1961, Jane Jacobs in *The Death and Life of Great American Cities* expressed the view that 'Old buildings are the shelter which is necessary and valuable to the district for many varieties of muddling, low and no yield diversity.' Keeping old fabric intact by preventing redevelopment may be a means of supporting communities and enabling them to survive rather than allowing so-called investment in districts, which, in the end, may remove both the fabric and the heart of communities. This is one of the social dangers of concepts such as facadism. With facadism comes the notion that the character of areas will be protected by preserving traditional street scenes and that by modernising the built environment behind facades both social and economic benefits will ensue. Sadly, the ensuing rehabilitation may contribute to damage to the social fabric of areas and the break-up of communities, as former populations are possibly displaced by higher income groups who can afford the cost of occupying the new accommodation and small businesses, once interdependent on one another and the local population, find that they can no longer survive and are forced to move out.

During the 1980s the concept of 'gentrification' emerged as an important factor in relation to such social and economic change within urban areas.[3] Originally the term gentrification was associated with a process whereby areas of historic working-class and derelict housing were rehabilitated and, as a result, were transformed into middle-class neighbourhoods. However, latterly, the term has been linked to a much wider urban renaissance prevalent throughout Europe and North America wherein the historical qualities of townscapes of commercial and early industrial capitalism, with their inherent tourism potential and cultural significance, have been promoted and marketed leading to large-scale social and economic change.

Residential rehabilitation is, therefore, only one facet of gentrification. It is a process which embraces commercial and industrial heritage generally and may typically involve, for example, the re-use of obsolete industrial buildings or the redevelopment of former dockland waterfronts to accommodate 'intellectualised' entertainment facilities, modern offices and 'trendy' bars and restaurants. A major driving force behind the transformation of many urban areas – which once either housed, or employed the working classes – has been the vast post-war expansion of the middle classes who have possessed the economic prosperity to move into and alter urban areas to suit their wants and needs. It is possible to detect a correlation between gentrification, the process of urban conservation and the concept of facadism. For instance, having been attracted by the potential of run-down or obsolete historic urban areas to provide good quality environments, the gentrifiers are likely to support the process of urban conservation to the extent that it protects what are perceived to be the most valued qualities. However, the gentrifiers will also bring pressures for physical change. They may, for instance, demand new retail, recreational or entertainment facilities in the locality. There is, therefore, a tension between the desire of the gentrifiers to protect some of the historic qualities of an area and their other demands for new facilities. Enter the various manifestations of facadism as possible development solutions, which offer the option of new facilities behind a historic cloak!

Facadism may not always be associated with dramatic social upheaval in central urban areas and in some cases local populations may welcome the changes that facadism can bring. For example, because facadism allows developers flexibility in remodelling historic fabric, it may provide the key to bringing investment into depressed parts of towns and cities and the possible social benefits of employment opportunities and improved facilities. There is no guarantee, of course, that local people will benefit from new employment opportunities. Employers may prefer to relocate their existing work-forces or recruit from outside the locality. Local shops may benefit from trade generated by the work-forces. Also the restoration or rebuilding of existing facades as part of redevelopment schemes may boost confidence in depressed areas and act as catalysts for further investment. If local people see street scenes being upgraded they may feel encouraged to remain in the area and become more socially active in working for more benefits.

The value of character, 'sense of place' and meaning in the built environment in terms of the quality of life (see Chapter 6, 'Facadism and Urban Conservation') illustrates a particularly relevant aspect of facadism to an examination of its social implications. For many citizens and visitors of towns and cities throughout the world, the facades of buildings will be the only significant part, this being the only element they ever experience. So many buildings are closed to the majority of people and the street is their most immediate behavioural environment. The permanence of this immediate environment may be a strong social value in a local neighbourhood and, in the event of irresistible development pressures for change, facadism may offer the only realistic solution to protect these values.

The educational role of historic buildings is another important social consideration. In addition to their associations with famous people and events, old buildings provide vital evidence of the way settlements have evolved and how architectural styles and building techniques have developed. Retained facades may well adequately remind us of specific historical events or particular persons associated with a building or place, perhaps in the same way that a sculpture might. Also, different architectural styles may be depicted on facades even if the interiors of the buildings have been demolished. Indeed, in the case of some historic buildings the interiors were utilitarian and somewhat mundane (e.g. Atrium Court, Glasgow, referred to on numerous occasions in this book) and the facades constituted the only architecturally valuable element worth preserving as an example of a design style. However, great care is required to avoid destroying vital information or artefacts, behind facades, because in some cases, facades only relay part of the story of buildings and places. Many eighteenth- and nineteenth-century facades are, for example, much more recent than the fabric behind, which often dates back to the sixteenth and seventeenth centuries. Features such as fireplaces or fine plasterwork, rare examples of plan forms or approaches to building construction may be of more value in an educational sense than the exterior of the building. Obviously, some artefacts can be removed and preserved elsewhere, but their true significance may be eroded in a different setting. Details may also be photographed or recorded in other ways, but how much more meaningful they are when experienced in reality.

ECONOMIC PERSPECTIVES

It is difficult to generalise about the economic implications of facadism in terms of the development process, because each case involves its own complex range of circumstances, financial constraints, costs and benefits. However, it is possible to highlight the ways in which facadism may influence the viability of development schemes and the use of local authority resources and also its economic impact on local communities.

In analysing the impact of facadism on the viability of development schemes I have considered three main elements in the development process:

(i) project planning and negotiations;
(ii) the construction stage;
(iii) marketing the scheme.

One of the most crucial factors in terms of assessing the viability of development projects is time. Within the life of development schemes numerous phases can be identified such as feasibility and viability studies, negotiations with planning authorities, site assembly, obtaining finance, construction and occupation or marketing and, in each phase, project planners must make decisions on the most cost-effective use of time. These decisions are fundamentally important to the financial success of projects, given the costs of obtaining funding for developments and the opportunity costs of not realising income from the sale or letting of completed schemes. Delays in any of the above phases result in higher costs, reduced developers' profits and, consequently, threats to the viability of projects. Also unforeseen delays could mean that by the time the development is completed the forecasted market trends have changed adversely.

When dealing with the historic environment in the UK, negotiations with planning authorities are of great significance in the earlier stages of development projects because of the potential for extensive delays in obtaining the necessary permissions. The growing power of the conservation lobby within the planning system in Britain since the mid-1970s has been such that negotiations over proposed development schemes have become complex time-consuming matters, regularly involving inputs from numerous different interests and the resolution of conflicting aspirations. It is often not simply a question of developers and local authority planning officers agreeing on proposals that meet their respective objectives. Where officers' recommendations have to be ratified by elected members it is possible that professional advice will be disregarded, leading to schemes having to be renegotiated. English Heritage and national amenity groups such as the Society for the Protection of Ancient Buildings (SPAB), the Royal Fine Art Commission and the Georgian Group exert considerable power over decision-making in terms of the built heritage. At the local level, conservation pressure groups, advisory committees and also influential members of the public can significantly affect development decisions. In assessing the feasibility and viability of projects, development interests therefore need to build in strong contingencies against delays in negotiating permissions.

This is particularly so where facadism is proposed, given the volatile nature of the debate surrounding the subject. My research reveals that a significant proportion of representatives of national conservation amenity and pressure groups and conservation officers within local planning authorities regard facadism as generally undesirable or as a last resort preferable to complete demolition of historic fabric. Development schemes proposing facadism are therefore often vigorously opposed by conservationists and may involve developers in expending large amounts of resources in justifying their aspirations. For example, it may be necessary to commission extensive structural surveys to demonstrate that buildings are beyond economic refurbishment and detailed market research to illustrate that the only realistic approaches to protecting the character of particular townscapes are either facadism or contextual new building. This, of course, generates the costs of professional fees in addition to the time costs of delaying the development to allow for the completion of surveys.

Having recognised the potential delays that may be incurred by development interests through adopting facadism as a chosen option, it should also be said that the concept of either retaining or rebuilding the facades and outer shells of some historic buildings may provide the key to resolving conflicts and releasing the development potential of urban sites. Where there is a lack of confidence within planning systems that contemporary architecture can satisfactorily protect or enhance the character of townscapes, but an equal unwillingness among investors to refurbish existing buildings, facadism may provide a viable compromise solution. Alternative strategies for development interests may lead, for instance, to lengthy and costly futile appeal cases and the eventual abandonment of proposals.

In the spring of 1992, while in Edinburgh, I discovered a development taking place at St Andrew Square, which is within the central part of the city known as the New Town. The site owners, Scottish Provident Institution, were implementing an approved scheme to expand their existing headquarters at No. 6 St Andrew Square – a 1960s building regarded by many as an exemplary building of the period – which involved the retention of the facades of Nos 4 and 5 and the front rooms of No. 4. Both buildings were listed (Scottish category B). The Italianate commercial palazzo, No. 4 St Andrew Square, by the architect John McLachlan, dates from 1883. As well as the handsome facade, it possesses a good-quality interior. The only discordant feature of the building is the fact that it is dwarfed in scale by the adjoining buildings, Nos 3 and 5. The latter building's facade is a somewhat austere example of Edwardian design, and its main townscape contribution is to provide a continuity of period architecture in the square (see Plate Section, xi). Its interior, being highly utilitarian, had little architectural merit. Negotiations on behalf of the site owners with the District Council, concerning the possibility of redeveloping the sites of Nos 4 and 5, had taken place in 1987 when the local authority planning department appeared to favour the retention of the majority of No. 4 but accepted the demolition and redevelopment of No. 5. The negotiations then progressed for two years, during which time the site owners appointed an architectural consultant, Roland Wedgwood, who produced an impressive contemporary scheme for the demolition and redevelopment of both Nos 4 and 5. This scheme was recommended for approval to elected members by the Director of Planning of Edinburgh District Council and was generally supported by the Royal Fine Art Commission for Scotland and an influential local amenity group, the Cockburn Association. Two other conservation groups, namely the Scottish Civic Trust and the Architectural Heritage Society of Scotland, expressed reservations about the proposals. The contemporary scheme devised by Roland Wedgwood featured a design concept which would have brought greater symmetry to the southern side of the square, to be achieved by the new building's curved facade being centred on the north–south axis of the square and symmetrical arrangement of the design features of the facade. The scheme also retained a reference to there originally being two building plots, improved the relationship of buildings on this frontage in terms of scale and provided a more sympathetic integration of Scottish Provident's existing headquarters. Despite the recommendation of the Director of Planning that the proposals be approved in principle, the elected members could not agree to the demolition of the listed buildings, feeling that this 'would affect adversely the character of the New Town Conservation Area'. A subsequent appeal by the site owners, against the refusal of planning permission and listed building consent, to the Scottish Office was dismissed.

The lack of confidence among the decision-makers that the contemporary design could adequately replace the listed buildings has led to the site owners finding in facadism a solution to their need for additional accommodation at their headquarters. Scottish Provident's consultation document (1987) outlining their requirements for expansion to meet the demands of the 1990s at St Andrew Square, stressed the need for 'flexible open plan spaces of sufficient size within the existing complex'. The achievement of consistent floor levels to facilitate the satisfactory installation of computer technology, data link services and air conditioning was also noted as an important factor. An appraisal of the space that could be provided by the existing buildings at 4 and 5 St Andrew Square was also included in the consultation document. It was concluded that floor levels would be at variance with one another and the cellular nature of the existing buildings would not provide the flexible types of space required. Also the costs of complying with building regulations and creating 'the necessary ease of circulation would be disproportionately expensive in relation to the floor areas achieved'. Finally, the financial constraints on the project were highlighted in the document and the point was made that any scheme retaining the existing buildings would result in less usable floor space, costly increases in circulation space and a consequent reduction in funds available to achieve the required quality of building. Although Scottish Provident's costs are likely to be greater by having to create their new office space within the constraints of the retained facades of Nos 4 and 5 St Andrew Square and the front rooms of No. 4, rather than having the flexibility of a completely new building, the removal of the interior of No. 5 provides a compromise solution, which one assumes was a viable option, as the development has since proceeded. Probably the only realistic alternative course of action open to Scottish Provident, and one with costly implications, would be partial or complete relocation either within Edinburgh or elsewhere. Given that the contemporary expansion scheme designed by Roland Wedgwood was of such high visual quality and still failed to convince the decision-makers that the demolition of the listed buildings was justified, it is unlikely that, by pursuing the option of a completely new building further, the site owners would have been able to achieve their aspirations. Facadism, in this case, enabled them to bring about at least a viable and workable (if not, in Scottish Providence's terms, ideal) solution to the conflict.

While considerable opposition exists among representatives of local planning authorities to the principle of facadism, its continual rejection based on dogmatic conservation principles may lead on occasions to extensive costs and demands on public funds. For example, a great deal of officer time and conservation budget funds may be spent on the futile negotiation of alternative less realistic options instead of organising and financing the repair of the most valuable historic fabric. In cases of derelict historic buildings, local authorities face the dilemma that long-term campaigns to save as much of the historic fabric as possible, because of ongoing deterioration of the fabric, may eventually result in the loss of the whole building, whereas accepting facadism may at least save important street frontages. Although it is possible – again at considerable financial cost – to secure and protect historic fabric to a certain extent from further damage, complete eradication of deterioration is impracticable, especially where severe rot has taken hold in the fabric. An awareness of the above economic realities is essential among decision-makers in the public sector to avoid waste of resources. Facadism may not, in many cases, be an ideal option for planning officers, amenity groups or elected members, but it may represent the only cost-effective way forward.

Some of the points mentioned above are illustrated in the case of Nos 1–6 Brunswick Square in Bristol, where ten years of negotiations between the local planning authority and a number of prospective developers almost ended in the loss of the southern frontage of a formal Georgian square. In 1984, a development scheme involving the repair and retention of the original facade and rebuilding of the side elevations was eventually implemented. Between 1974 and 1981 the local planning authority had attempted to achieve the preservation of much of the form of the original late eighteenth-century terrace, but accepted at the end of this period that, of the original fabric, only the facade could realistically be preserved. Since the facade, which forms the southern frontage to an elegant Georgian square, was saved and the side elevations rebuilt, the essence of the formal townscape composition is still in evidence and it could be argued that the main conservation priority was achieved. However, by the time the development was commenced, the facade was in a critical structural condition and in danger of collapse. Consequently, it required considerable repair, which was grant-aided from the Council's Conservation Programme budget, to enable its retention. Nothing remains of the original interior and its form was not reflected in the new development.

Ironically, had the Council's attitude towards its historic environment been the same throughout the 1970s, more of the original fabric could have been retained and the costs expended over a decade in trying to achieve an appropriate conservation solution could have been saved. In 1972, and again in 1973, a firm of developers had approached the planning authority with a view to accommodating Nos 1–6 Brunswick Square in a development comprising small office suites. This would have afforded an opportunity to negotiate the retention of the original internal form of the buildings. At the time, there was little commitment within the authority towards conserving Bristol's built heritage and detailed planning permission had already been granted for a proposed 'Modern' scheme to cover the southern frontage to Brunswick Square. The Council were awaiting the results of a challenge in the High Court against the Secretary of State's decision to overrule his appointed Inspector and refuse listed building consent to demolish the buildings. Little encouragement was given to the proposals to retain the existing buildings and negotiations ceased. In effect, the Secretary of State's intervention in the determination of the application for listed building consent saved the buildings. The Brunswick Square case also demonstrates how schemes can flounder because of extreme changes in 'the market'. By 1978, the Bristol Conservation Programme had been set up with the Council's financial commitment being matched by the then Historic Buildings Council. The availability of grant-aid incentives to developers had enabled the planning authority to negotiate an office development scheme which showed internal walls, fireplaces and the rear elevation of Nos 1–6 Brunswick Square retained. However, economic circumstances at the time, which severely suppressed demand for office accommodation, led to the scheme being abandoned. Even the offer of 50 per cent grant aid from the public purse was insufficient to encourage implementation of the scheme.

In cases of facadism, the costs of preserving or rebuilding historic facades, and creating replicas, are important elements in the financial equation of development projects. To the construction costs for the new build elements must be added allowances for a range of other measures. First, the cost of providing an elaborate scaffolding system to safely shore up an old and possibly structurally delicate facade during the construction of a new building behind, is likely to be a significant element. Even though this will be offset, of course, if the old facade is successfully retained, by the fact that a new facade does not have to be constructed, the inclusion in financial appraisals of a contingency plan to cover the collapse of the old facade once the rest of the former building has been removed, is advisable, given the inevitable uncertainty prior to work commencing on site about the precise structural condition of the facade. In many cases, historic facades will require extensive refurbishment and cleaning as part of their preservation, to make good age-affected materials. Environmental damage will often have weakened a facade's structure and eroded architectural features. In particular, in the case of many Georgian buildings, because the lintels, roof and floor joists were timber, the loss of lead gutters and flashing may have caused water penetration, followed by deterioration and structural failure in the timber members and consequent cracking in surrounding stone or brick fabric. Further repairs may also be required to make good any damage caused by having to tie the facade to the scaffolding structure.

Where old facades are to be preserved and attached to new buildings, particularly sophisticated structural techniques are required, which need to be carefully appraised in terms of cost. The adopted measures must take account of the fact that the new building and the facade will behave differently in terms of movement. Also, the creation of an exact replica of a former facade may require elaborate detailing and ornamentation and the use of particular natural materials, which will be expensive elements in the scheme.

The financial costs to developers of facadism may in some cases be offset by grant aid from local authorities and English Heritage. While English Heritage is generally opposed to grant-aiding schemes which only preserve the facades of listed buildings, exceptions are made in certain circumstances where facades are regarded as vital townscape elements in their own right. For example, English Heritage and Bristol City Council paid grants of approximately £100,000 during the 1980s towards retaining the facade of Nos 1–6 Brunswick Square, out of a total project cost of around £4 million.

The 'critical path' technique of appraising development projects involves an identification of the key factors, stages and events throughout the project which are crucial to the completion of a viable scheme. I discussed the question of whether or not supporting a historic facade during a redevelopment scheme would be likely to affect the critical path of the project with both the structural engineer and project architect for the CIS development in Bristol which resulted in Brunel House (see Chapter 7, 'Why Facadism?', Case Study 2), referred to on numerous occasions elsewhere in this book. Both expressed the sentiment that shoring up a facade from the outside of a site would be unlikely to make much impact on the scheme's critical path. The scaffolding fully supports the facade until it is tied into the new building behind, which is structurally independent. To try and ensure the avoidance of problems in the case of Brunel House, monitoring equipment was set up to assess continually the structural stability of the facade during the construction of the new building, and the scaffolding was designed to accept approximately 150mm of lateral movement in the facade. Greater difficulties and financial costs are generated in schemes where facades have to be supported from within the site, because of the increased inconvenience and constraints to contractors.

It seems that in Britain, the conservative attitudes of funding institutions may, in some circumstances, encourage developers dealing with the historic environment to propose various forms of facadism rather than the refurbishment of obsolete old buildings. The providers of funding must be satisfied about the security of proposed investments and will often prefer schemes resulting in the creation of new buildings which are perceived as involving far less financial risk than the business of restoring and adapting old fabric. Forms of facadism such as facade preservation and replication in effect result in what are essentially new buildings.

A further financial element which may, on occasions, influence development decisions involving the historic built environment, are the relevant tax laws of a country.[4] In the British Isles, according to Tugnutt and Robertson (1987: 86) 'the vagaries of the tax system, and in particular Value Added Tax, can provide a significant inducement to demolish and build anew'. They link this to proposals within historic townscapes to construct new buildings with replica facades.[5]

In general, in mixed economies, development schemes must be marketable to be viable. If property investors and developers are to initiate development, they require assurances that the buildings created will produce adequate returns on their investments, either through capital values or rental income. Where new development is intended for occupation by the developer, or has been pre-let before construction, the design requirements will be clear. For example, in the case of Nos 4 and 5 St Andrew Square, Edinburgh, Scottish Provident Institution, the site owners, specifically required flexible open-plan accommodation with consistent floor levels within which to expand their headquarters. Alternatively, the development may be purely speculative, in which case it will need to appeal to high-demand sectors of the property market.

The management of historic urban areas involves balancing the above economic considerations with aspirations for preserving and enhancing the character of the local environment. In some cases, the typical cellular structure of particular historic buildings may not facilitate viable conversion schemes to meet market demands. For example, in the office and retail sectors, investors may feel that, rather than refurbishing existing buildings – which have, for example, interiors comprising generally small spaces and regular horizontal subdivisions – or reflecting their form in a new development, it is more prudent economically to create large, adaptable, open-plan spaces within which clients are given the flexibility to arrange or subdivide the accommodation to their requirements. This implies that historic urban areas must sometimes adapt to reflect social and economic change. The attraction of facadism is the flexibility it provides for creating more widely marketable and economically viable accommodation while keeping traditional streetscapes intact. A visit to most of Britain's historic towns and cities will demonstrate that such an approach has often proved irresistible. Numerous examples are to be found where facades have been preserved or rebuilt, behind which old, small-scale spaces have been replaced by modern open-plan accommodation. These new buildings, because of their greater flexibility and versatility, are often the key to attracting major investment institutions. A consequence of the influx of larger-scale national or multinational organisations is likely to be the displacement of smaller and often locally based businesses, many of which will have helped to characterise particular areas for many years and also provided vital services to surrounding communities. Clearly, when seen in such a light, facadism may be considered a concept which encourages fundamental changes in the economic structure of historic urban areas. Some communities will feel that they have benefited from the injection of new investment; others may consider that they have lost their soul as the anytown image of multinationality stands glittering on the grave of local enterprise.

Finally, the image-making role of facadism should not be overlooked. While it may be an important factor in many development schemes to create modern, flexible accommodation, the exterior image of new buildings is also an important marketing consideration. For instance, some potential occupiers may be more attracted by the image generated by a traditional facade than one of contemporary design. Also, it may be beneficial to a company's image to be seen to be contributing to conservation initiatives in an area by accommodating its new operations within the traditional envelope of a preserved or facsimile facade. Judging by the enthusiastic public response, this appears to have been the case with Marks and Spencer's city centre retail development at Exeter in the early 1980s, where part of the store envelope was formed by rebuilt historic facades. On the evidence that so many developers have received planning permission throughout the UK during the 1980s and early 1990s to create modern central-area shopping precincts within the cloak of either preserved historic facades or new versions evoking the traditional image, I would expect considerable public support to be revealed in a national survey of attitudes towards such an approach to urban remodelling.

6

FACADISM AND URBAN CONSERVATION

During the last twenty years, in Britain and many parts of Europe, the process of conserving the urban-built environment has become a most influential aspect of the development process. The post-war attempts to divorce the past and create a 'brave new world' through comprehensive redevelopment schemes and a dogmatic adherence to Modernist principles in building design have given way to a new consensus, within which there exists a strong will to retain familiar historic townscapes and knit the past and present together through more contextual forms of new development. Traditional materials, craftsmanship and ornamentation, reviled by the Modern movement, are respected once more. The motivations which fuel the desire to conserve the built environment are diverse and embrace philosophical, psychological, aesthetic, social, political and economic themes.

Consider, for instance, the important contribution of historic environments to the quality of many people's lives and the impact of built surroundings on the state of the human mind. Buildings and places of great architectural quality provide wonderful aesthetic experiences. Different approaches to conservation of the built environment will affect local communities in different ways and create a range of economic costs and benefits. The conservation process is also subject to powerful influences from governing agencies, pressure groups and market interests. Development decisions concerning historic environments are, therefore, the result of a highly complex system of interactions and power struggles. Facadism is a concept at the interface where pressures for urban change come into contact with the processes of protecting the built heritage. From one standpoint, facadism can be seen as a compromise solution which resolves certain power struggles. However, numerous misgivings exist about its validity in terms of achieving urban conservation objectives. In this chapter the many important implications raised in this debate will be examined.

To assess these implications, it is necessary to form a conception of what the process of urban conservation is in essence and to highlight the motivations which fuel the desire to conserve the built environment. Although urban conservation is a somewhat elusive subject, a number of prominent themes and principles can be identified which help to define its essence, and so the implications posed by facadism may be dealt with in the light of these. We may look to a number of sources in identifying these themes.

One such source is what can be termed an 'official' view of the process of urban conservation set within a contextual framework of town planning law, policy and practice. In England, the underlying themes identified from such a perspective are of fundamental significance to an examination of the conservation–facadism relationship, given the influence of centralised planning policy over decisions and outcomes in the urban conservation process, and so, for the purposes of this book, this country's policy framework provides the focus. It must also be appreciated that urban conservation is not a self-contained activity but part of a much larger urban process. It can be seen as an organising principle[1] and form of urban management which is embraced by overlapping and interlinking physical, social, economic and political factors. It is therefore inextricably linked to the land development process within which there occurs a private/public sector interface and a complex interaction takes place between various actors and agencies constituting a range of interests. Conservation is now a highly significant consideration in urban design[2] and a study carried out by Gleave and Wheeler (1988) demonstrated the increasing importance of an awareness of the complexities of the development process in urban design matters generally. They concluded that a 'traditional understanding of urban design as a scientific or technical discipline related in many cases to architectural objectives, is no longer adequate'.[3]

This organising principle is part of the process that has become known as 'heritage planning',[4] which combines the concerns for buildings, relics and artefacts of the past with the realisation that they represent a valuable resource which can support present-day urban functions through, for example, tourism and the provision of highly attractive heritage environments. Consequently, urban managers and planners increasingly utilise the heritage environment as part of the mechanisms of urban revitalisation and renewal.

Although the themes which can be identified from examining 'official' urban conservation policy and practice are of fundamental significance in attempting to identify the spirit or essence of urban conservation, they cannot be considered definitive. The policies are merely guiding principles which represent governing agencies' views of how urban conservation should be practised within the land development process. An exploration of the subject reveals other ideas which extend, challenge or question 'official' policies. Decisions taken at the local level may not always reflect views at the centre. Alternative perspectives may be found through a review of appropriate literature. Other views emerge from the individual's own perception of what the process of urban conservation should be. Having arrived at a conception of what urban conservation is in essence, and deliberated on the extent to which facadism reflects the conservation ethos, we obtain further enlightenment by considering the motivations which fuel the different desires to conserve the built environment and colour the various perspectives on the subject. An exploration of these motivations reaches the heart of the matter. By considering why we seek to conserve the built environment we may examine how far facadism takes us towards achieving our objectives.

'GREEN' PERSPECTIVE

Conservation, in general terms, is a process involving the husbanding of the world's natural resources in an environmentally sustainable manner and maintaining nature's balance for the wellbeing of this and future generations. In the context of the built environment, such a philosophy can be translated into a number of development principles. A fundamental tenet of conservation must be restraint in the use of non-sustainable natural building materials. In some situations, where visual impact is not a critical issue, reconstituted or artificial materials may suffice instead of natural resources. Also, the re-use or adaptation of buildings rather than demolition and redevelopment enables the reduction of demand for new extraction of materials such as stone and slate. Dobby (1978) identifies the ecological crisis, highlighted by European Conservation Year in 1970, as a significant influence on the growth of a desire to conserve the built environment in that the husbanding of finite resources became an imperative and encouraged the move away from the attitude of wasteful clearance prevalent in the 1950s and 1960s. It is a sobering thought that as many as 8,000 listed buildings were demolished between 1957 and 1977 in the UK, many through local authority road schemes.

Another important urban conservation objective ought to be the reduction of carbon dioxide emissions and other pollutants through, for example, the creation of energy-efficient buildings and control of vehicular traffic movements. In Australia, state codes for residential development nowadays highlight the importance of energy efficiency in design. Scandinavia is very advanced in such practices and, in recent years, several other central European countries have been placing greater emphasis on environmentally sound design. Sadly, Britain is not as advanced as many of its European neighbours in this respect[5] and, in terms of the town planning system, central government has not yet demonstrated a commitment to this aspect of urban conservation. However, the cause is being pioneered at the local level. In the county of Somerset, for example, a local plan review document at the deposit stage of the statutory process, produced by Mendip District Council in 1991 and relating to the Glastonbury and Street area, contains a conservation strategy – for which I must claim responsibility, having prepared this part of the document – wherein a central aim is to encourage 'developers to plan with the environment generally in mind and to think not only about aesthetic criteria but also whether or not their proposals help to sustain the environment or result in the efficient use of resources'.[6]

In relation to the environmental effects of motor traffic, many towns and cities in the UK are facing up to the issues and introducing a range of measures such as traffic-calming schemes, pedestrian precincts and cycleways. Many such measures come about following policy statements in local plans and conservation programmes. Inevitably, however, central government restrictions on local authority expenditure places the emphasis for implementing such proposals on private-sector interests within the development process.

Is facadism of any significance in terms of this 'green' perspective of urban conservation? In relation to the use of non-sustainable natural resources, the prevention of the complete demolition of historic buildings, by the retention or rebuilding of their facades utilising as much as possible of the original fabric, can be seen as a valuable act of conservation. Within historic environments, the textures, tones, colours and weathering of the natural materials used on the facades and roofscapes of buildings are often fundamental elements in the special character, aesthetic quality and local identity and, clearly, one of the established principles of urban conservation is the preservation of such characteristics. In particular situations, where historic areas are under irresistible pressure to evolve, the retention or rebuilding of facades is a vital factor in preserving the sensitive balance between the various visual elements which contribute to a sense of place. Also, from a resource conservation perspective, the fact that a new facade does not have to be constructed, and the possibility is provided of using modern sustainable materials behind the facade, means that demand for new natural materials is probably reduced. There are, of course, other conservation options. Obsolete historic buildings may be replaced by new buildings with replica facades or, alternatively, other contextual forms of new development. In some situations, the use of reconstituted or artificial materials in new development may be contemplated, and such a course of action would reduce demand for new natural materials. However, such materials possess different qualities of tone, texture and colour and have different weathering properties; therefore, to be able to achieve the necessary architectural and aesthetic quality in the replacement development, the use of appropriate natural materials will, in many instances, be essential, at least on the facades of new buildings. Even then, unless recycled and weathered natural materials are used, the effect of the changes will be felt because the patina of age, often an essential ingredient in character, takes a long time to develop. It is possible to recycle some natural materials after demolition, but large quantities are inevitably damaged and lost. Consequently, the constant process of knocking down historic buildings for redevelopment is wasteful and, given a desire to conserve the unique character of historic townscapes, eventually unsustainable. In the light of the above thoughts, if urban conservation is to embrace environmental management in its broadest sense, concepts such as facadism have an important part to play.

Another aspect to be considered here is the opportunity facadism provides to create much more energy-efficient buildings within historic urban areas. The designer has basically a clear site or empty shell of a building within which to incorporate environmentally sound techniques, which is potentially much easier than adapting an existing historic building with its inherent constraints. The outcome, therefore, can be an unchanged street facade, with a more environmentally sustainable building behind.

The management of vehicular traffic in urban areas is an important issue in terms of environmental conservation. Consider, for example, matters such as atmospheric pollution and chemical attack to the built fabric as a result of vehicle exhaust emissions, the damage to buildings caused by vehicle impact or vibration and the undesirable effects of large numbers of vehicles on the visual quality of historic areas. Somehow a satisfactory balance has to be reached between protecting and enhancing the environment and successfully managing traffic to enable places to function effectively and survive both socially and economically. In some situations there may be a role for facadism in helping to achieve these objectives. By enabling the radical remodelling of buildings, or groups of buildings, behind historic frontages, it may provide opportunities to create new access and parking arrangements, or street patterns, and through this achieve numerous conservation objectives. For example, facadism may allow the provision of a new parking and servicing arrangement to the rear of a historic shopping street frontage, which in turn facilitates improvements to the shopping street, such as changing priorities in favour of pedestrians, cyclists and people with mobility difficulties and also the reduction of fumes, vibration, noise and the unpleasant visual clutter of vehicles and traffic signs. In such a situation there is not only the potential to create a more environmentally friendly milieu for shoppers within a familiar street scene, but also to provide a more efficient servicing operation for traders, improve parking facilities and reduce the threat of damage and destruction to the remaining historic fabric. While there are numerous benefits to be gained from such a course of action, facadism is a drastic approach to dealing with historic townscape, and there are circumstances where it would be totally inappropriate: for example, where the character of an area relies heavily on the preservation of historic street patterns and whole buildings as opposed to just frontages. More preventive measures rather than the harsh surgery of facadism must then be employed. This may well involve much more extensive exclusion of motor traffic from historic areas. If societies wish to truly preserve the special character of their historical and architectural heritage, there must also be a realignment of views towards motor vehicles and an acceptance that in many situations private motor cars cannot be accommodated or that large delivery vehicles are not able to transport goods from the warehouse direct to the shop door. In discussing this point a vital social dimension should not be overlooked. Environmental enhancement is an important element of conservation and should include making the conserved urban area as widely accessible as possible to all groups of the community. Therefore, where traffic is excluded from historic areas, some valid exceptions for public service vehicles and disadvantaged groups, such as the disabled, should be made.

'THE SYSTEM' IN ENGLAND

The notion of an environmentally sustainable approach to development should be an important facet of urban conservation and there are signs that the UK may now be starting to follow examples of environmentally sound building practice evident in parts of Europe, Australia and the USA. However, the established 'official' approach to urban conservation in England has so far concentrated on three main facets:

1 The preservation of ancient monuments, under the direct control of the Secretary of State, supported by English Heritage.

2 The listing and protection of buildings of special architectural and/or historic interest. It is an offence under planning law to demolish, or undertake work which affects the character of listed buildings without first obtaining the consent of the local planning authority and, in some cases, the Secretary of State.
3 The preservation and enhancement of the character and appearance of conservation areas designated by local authorities. The purpose of conservation areas is to protect and enhance the character and appearance of parts of the built environment that have a recognised value either architecturally or historically, or both, even if individual buildings may not warrant the protection of listing. Designation is often, therefore, a recognition of the group value of buildings and the special qualities of their settings.

It was not until the early 1970s in England that the state showed any form of concerted commitment to intervene in the development process with a view to conserving the built environment. It is true, of course, that the notion of protecting valuable parts of the built environment did not suddenly spring up at the end of the 1960s. Even in the nineteenth century William Morris was promoting the principle of preserving ancient monuments, and the system of listing buildings dates back to 1944. However, prior to legislative reforms during the late 1960s and early 1970s, an emphasis on protecting individual buildings and a reliance placed on what were largely inadequate statutory powers proved ineffective in conserving many areas of townscape which would probably now be regarded as valuable. As Burke points out, writing in 1976:

> Until recent years conservation was scarcely recognised as a subject for national government policy. Buildings and ground layout from the past survived fortuitously, largely on their own merits, and chiefly because they could continue to serve useful purposes.
>
> (Burke 1976: 117)

The situation changed considerably in 1972 when demolition of unlisted buildings in conservation areas was brought under planning control and the Secretary of State, on the advice of what was then the Historic Buildings Council, was empowered to make grants or loans with a view to preserving or enhancing the character and appearance of areas of townscape value.[7] This period saw the emergence of the notion of preserving or rebuilding facades as part of conservation initiatives, although it did not feature explicitly as a concept in central government advice. However, turning to the local policy perspective, I have found an example where facadism was being accepted as a valid approach to urban conservation in appropriate cases. In 1974, Bristol City Council designated the Portland Square Conservation Area, covering an elegant example of late Georgian expansion around the original city centre. The conservation area policy document of the time states: 'Where sites form an essential part of the street pattern, redevelopment should involve the rebuilding of facades so that the street pattern may be restored.'[8] Subsequently, a number of facades in the Portland Square locality and elsewhere in the city were retained or rebuilt, mostly as part of office redevelopment schemes.

Current advice being offered to local authorities in England by central government, and the powers available for conservation of the built environment, are outlined in Circular 8/87, which updates and consolidates previous advice (Department of the Environment 1987). Again, facadism is not mentioned specifically, but the contents of the Circular are most relevant to an analysis of the role of facadism in the urban conservation process: details are given of listing criteria; powers to deal with buildings in need of urgent repair; and financial matters. The listing criteria are important in identifying why certain buildings are selected for protection and why, in some cases, facadism may or may not be appropriate. Particularly relevant is the listing of some buildings for their group value within the townscape, where the emphasis is not on an individual building, but on the relationships between several buildings, their juxtapositions and their external features. It could, in some instances, be argued that the most important conservation objective will be the preservation of the facades of such buildings. This was a fundamental issue debated at the public inquiry in 1989 relating to proposals to demolish eight listed buildings at the 'Mappin and Webb site' in central London (see pp. 116–19). In this case the conservation lobby campaigned vigorously for the retention of the facades of the listed buildings, albeit without success in terms of winning the argument. It is of interest that, on this occasion, the Secretary of State rejected such an approach in favour of James Stirling's modern statement.

It is also interesting to note the view of English Heritage, the quango set up by the 1983 National Heritage Act, which is responsible for scheduling ancient monuments and listed buildings and providing grant aid towards their upkeep. The body is generally opposed to facadism as an approach to dealing with listed buildings:

> Our policy is to resist the demolition of listed buildings behind a facade since it is the whole building which is listed. The facade alone is less likely to be listed and the action may remove the building from the list. In this situation a grant would not be available. However, in a few cases where a building has been structurally unsound a grant has been offered to repair the facade and the rest of the building has been reconstructed in a more modern form.[9]

Cedar Lodge at Frenchay near Bristol, for example, was a property de-listed following its adaptation which involved the retention of just the facade.

The power to deal with buildings under threat because of neglect, outlined in Circular 8/87, is also relevant to considerations about facadism, as is the ability to finance preservation or restoration of complete buildings rather than just facades. Buildings that are allowed to become derelict through neglect may decay so badly inside that the only options are to demolish them completely or, perhaps, if structurally feasible, retain the facades because of their townscape importance. Measures are available to planning authorities to prevent deterioration of listed buildings or to compulsorily acquire them to secure their restoration. Such measures can prevent the need to resort to solutions such as facadism, but there are financial implications. English Heritage is, in effect, a centralised managing agency for funding urban conservation and demands on its resources through grant-aid applications far outweigh the amounts available; despite expansion during the 1970s and 1980s of a financial and legislative commitment by central government to conserving the built environment, great reliance is placed on local authority funding and the co-operation of the private sector to achieve conservation objectives.

Clearly, local authorities do not have the resources to save every historic building under threat. During economic recessions, such as in Britain in the 1970s and latterly in the late 1980s and 1990s, private-sector development activity tends to be curtailed, resulting both in the postponement of schemes which might otherwise have contributed to the achievement of conservation objectives, and in developers becoming more cautious and less willing to invest in costly conservation schemes. The legal provisions and policy statements do not eradicate the threats of decay to the historic built environment, and some cases of facadism can be seen as a symptom of the situation and a pragmatic response to conservation problems: a last-resort solution to prevent the complete loss of a historic building, or as the only economically feasible option available to retain the character of a streetscape. This is illustrated well in the case of Nos 1–6 Brunswick Square in Bristol, where the City Council, after ten years of negotiations with developers (1971–81) and offers of grant aid, were unable to secure anything more than the retention of the facade of the original buildings, which formed the southern side of a Georgian square. In 1978 a scheme was granted planning permission which preserved a considerable amount of the original buildings. However, market conditions were not conducive to the development taking place and, by 1981, the remaining fabric, including the facade, was in serious danger of collapse. A decision was then taken by the local authority to allow the construction of an open-plan office development with simply the original facade facing the Georgian square restored and retained.

The city of Chester's response to the financial demands of urban conservation is to levy a conservation charge on the local community, the funds for which are exclusively devoted to protecting the historic fabric. This type of initiative, if politically palatable, gives a local authority considerable leverage in dealing with historic buildings in terms of offering grant-aid incentives towards refurbishment or allowing compulsory purchase of threatened buildings. Also, an important aspect of successful urban conservation identified by Chester City Council is adequate monitoring of historic buildings to alert the Council to problems before situations become critical or beyond redemption.[10] Organising urban conservation on such a basis would appear to enable planning authorities to be more pro-active in managing the historic environment. It could, for instance, prevent the presentation by a developer of a *fait accompli* to an authority where a building has become so badly mutilated or decayed inside that the only options available are to salvage the existing facade, accept the construction of a replica with new building behind or allow a contemporary building.

With increased central government pressure on local authorities during the late 1980s and early 1990s to reduce expenditure, and the inevitable impact on conservation budgets, it is surprising, given the apparent weight of opinion within the planning profession against facadism as a valid approach to urban conservation,[11] not to find local authorities using other initiatives, such as adopting design guidance, to oppose facadism. My research for this book did not reveal one major city or historic town in Britain where such policy guidance existed. Perhaps this is indicative of an acceptance of the physical, social and economic realities that historic buildings and areas have to evolve as part of the conservation process and that sometimes change will be extensive. Perhaps too, there is an acceptance of the architectural dimension: that resisting a technique which involves conceptually divorcing the exterior of a building from its interior, on the grounds that it infringes some great architectural tradition or principle, is indefensible.

URBAN MANAGEMENT

The development process involving the historic and architectural heritage is an arena of conflict for competing interests. Not uncommonly, the demands of large commercial interests, who, for example, may be seeking to promote a national image or to create modern adaptable accommodation, will be at odds with the views of conservationists, who may, perhaps, favour the status quo or feel that the approach taken by the development interests in a particular situation would destroy the special character of either an individual building or even a whole locality. It is clear that a fundamental aspect of the process of urban conservation is the resolution of such conflicts brought about by the pressures for change within the historic built environment. Therefore, conservation cannot be a self-contained activity but is part of a much larger urban process. It can be seen as an organising principle and form of urban management, which is embraced by a range of physical, social, economic and political factors that often overlap and interlink within land development decision-making. The historic built environment has become a valuable resource to be utilised in the satisfaction of contemporary demands and needs and in the revitalisation and renewal of towns and cities. This aspect of urban management has become known as 'heritage planning'.

In England, Circular 8/87 reflects the urban management perspective in stating that 'conservation work should not be regarded as a separate facet of local authority activities' (Department of the Environment 1987). In 1984, Chester City Council's Conservation Review Study draft report indicated that conservation was a key element in the economic regeneration of the city and in the creation of a high quality of life for the residents. In Bristol, the Council's Conservation Programme, begun in 1979, was seen as an extension of a whole range of programmes to promote economic growth and improvement to the inner area of the city and never as an end in itself. Shrewsbury and Atcham Borough Council, in a publication entitled *Conservation and Tourism* (1981), highlighted an idea, that may often escape those with an excessively preservationist perspective of urban conservation, that conservation should not lead to town centres becoming mere showpieces but that they should serve a variety of functions.

Just as people have to adapt to survive in a rapidly changing world, if urban settlements are to continue as living communities they must always evolve physically, socially and economically, to reflect new trends. As part of a system of urban management, conservation must be integrated with other initiatives such as enabling people to meet their needs in terms of finding adequate housing, employment and obtaining necessary services. These initiatives will, potentially, have a significant effect upon urban systems, on the way they operate, and consequently upon the character and appearance of places. In this context, I find the following observation by Donald Insall illuminating:

> Conservation in Britain is an integral part of land-use planning . . . a guiding principle upon the processes of change. In this way a historic town can be kept 'alive' and not frozen like a museum piece. The art-historian philosophy is most dangerous – you cannot 'put-back' a town to its original state – what is the 'original' in a living town? What one *can* do is to incorporate sympathetically into it, and then enjoy those earlier human strata which are so precious to it.[12]

Facadism is an important concept in the context of urban change and economic regeneration in historic areas. One of the most significant points is that it allows greater flexibility in sympathetically incorporating new accommodation within traditional settings of high visual quality than does the conversion of existing fabric. At George Street in Edinburgh's New Town, for instance, radical alterations to a number of buildings have been sensitively achieved by building new office accommodation either behind existing facades or behind the facades and front rooms of the original buildings (see Plate Section, viii). In Glasgow, the beautiful Venetian-inspired facade of the Ca'D'Oro building (see Plate Section, ix and x) which had become obsolete, cloaks the rebuilt interior, which has given the building new life in the form of retail and office use.[13]

The urban management process of heritage planning is demand-led, treating the historic environment as a product to attract consumers with a hoped-for return which will contribute to the future maintenance of the environment and support of local economies. Marketing strategies need to adjust to the nature of demand. For example, full demand will suggest the need to maintain existing conditions. Latent demand will require a developmental strategy, faltering demand remarketing. A total lack of demand will require action to stimulate interest, whereas excessive demand may suggest the need to demarket a place (see Ashworth 1991). The flexibility of facadism in some situations provides a useful development option to consider in connection with marketing strategies. Facade preservation or replication in front of completely new buildings may enable urban managers to remarket areas and stimulate demand by retaining the historic atmosphere of streetscapes and, at the same time, changing the structure of the local economy by attracting new uses and occupiers. It may be felt, for instance, that it is desirable to create modern retailing accommodation to bring in a magnet store within a high-quality heritage setting. Alternatively, the aspiration may be to create a new residential area or leisure complex.

Let us briefly focus on a number of examples of what are, in essence, cases of facadism to be found in former economically redundant docklands in Bristol, Gloucester, Liverpool and London. I have referred elsewhere in this book to the significant contribution made by the Arnolfini Arts complex, once the Bush Warehouse in Bristol City dock, in the revitalisation of the quayside. Just across the waterway the shells of other former dock buildings accommodate the Watershed Arts and Media centre: shops, cafés, bars and an exhibition centre bringing life to what had become a desolate commercial wasteland in the very heart of the city. Albert Dock in Liverpool has also been previously mentioned, where the internal remodelling and re-use of the Grade I listed buildings has created a centre that has become an attraction of national significance. In the City of Gloucester and at St Catherine's Dock in London urban renewal has been brought about through redesigning the interiors of traditional dock buildings to house new uses while using the exterior to retain the visual dockside character.

A different approach may be to provide new facades to existing buildings in order to implement a marketing strategy. This is less common in the UK than facade preservation or replication. Two extreme examples of this type of facadism are to be found at Yakima (Wash.), USA. Here, 'The Train Mall' and 'Yesterday's Village' create heritage to dress up the contemporary leisure and shopping functions. Three historical elements dictated the type of heritage image created: the importance of the railway; the native Indian community; and the Wild West legends. In the case of 'The Train Mall', a set of wooden facades and a boardwalk have been added to a collection of railway coaches and wagons to accommodate a variety of restaurants and speciality shops. 'Yesterday's Village' is a converted warehouse, which has been decorated externally with a veneer representing the typical image of a Wild West town (see Ashworth 1991).

MORE THAN PRESERVATION

Conservation has a multi-dimensional nature. Although obviously inextricably linked to the past, it is a process which is concerned with the present and is also forward-looking. It is not simply a matter of restoring buildings to their original condition or preserving buildings as they are at present. There will, of course, be occasions where pure preservation of a building or group is essential in conservation terms, for example, because of the rarity value of the architectural features both inside and out, or importance of historical associations. However, as Burke indicates in his book *Townscapes*, written in 1976, there is a distinction between conservation and preservation:

> Preservation implies static protection, saving from decay keeping things as they are. Conservation means, or has come to mean, preserving purposefully; giving not merely continued existence, which often implies retaining or restoring the traditional appearance of buildings.

Burke's definition points to the creative and dynamic nature of conservation, which is a view echoed by Francis Tibbalds, who provided the following definition:

> Conservation to my mind is correctly treated as a creative approach to change and modernisation – a socially and economically viable alternative to indiscriminate demolition and redevelopment.[14]

Conservation clearly implies something more than the embalming of museum pieces as an end in itself and, as Tibbalds remarked, 'is part of a sensitive yet lively process of urban adaptation and renewal whose object is to improve all aspects of the quality of life'.[15]

Although in principle conservation is a process which attempts to accommodate pressures for change within historic urban areas, a major thrust of conservation practice is the resistance of the wholesale type of change that occurred in British towns and cities during the 1950s and 1960s. The wastefulness of demolishing buildings which are capable of being restored and reused is thus avoided, and continuity and stability within the built environment achieved. Owen observed that:

> Urban conservation is the planned reduction of the rate of physical change in those parts of settlements that have long-term social environmental and historic value whilst continuing to accommodate people's changing and unchanging needs as expressed through their activities.[16]

When urban change takes place on a comprehensive scale it can be a painful process. During the post-war years British society looked for a 'brave new world' after the experiences of the 1930s Depression followed by the horrors of war. To facilitate new social and economic patterns and replace parts of a built environment ravaged by German bombs, many town and city centres were virtually rebuilt. In retrospect many people came to view the changes as brutal and uncompromising actions that had created unfamiliar poor-quality townscapes and destroyed communities. By the 1970s a strong thrust of opinion was rising against further sweeping change and the momentum of the conservation movement, which has been maintained ever since, was established.

Facadism, in some circumstances, can be part of the dynamic process of conservation. Applied sensitively it preserves certain townscape characteristics, while allowing improvements to other aspects of the quality of life. In the case of Brunel House in Bristol, for example, the most dominant and vital townscape feature of the original building, the facade, has been preserved and good quality office accommodation constructed behind. It is unlikely that such light and airy work-space with generally pleasant views to the outside world could have been achieved by converting the interior of the former building. The same principle applies in the case of Atrium Court, Glasgow, where high-quality modern office accommodation has been created within three sides of the former building (see Plate Section, i, ii and iii).

There is an argument, however, put forward by SAVE that 'Facadism has not only destroyed interiors of value and wasted precious resources, it has deprived people of the pleasure of working in spaces of distinct character and quality.'[17] This idea could be taken further to include other non-work uses such as dwellings, buildings for leisure, shopping environments and so on. Clearly, these are subjective considerations and revolve around people's tastes. Some will prefer more modern, functional surroundings while others will feel more in tune with a more traditional setting. The challenge of conservation is to ensure that high-quality interiors are not lost but given continuing life through appropriate uses. The charm and character of some conservation areas derive from the fact that the buildings have survived and remained economically viable for many years intact without large-scale modernisation. A recent visit to Dartmouth, the ancient riverside town in Devon, accentuated this for me. In such places, the mists of history have not been swept away by the winds of change, nostalgia is rife but so too is a timelessness and a reassuring sense of solid endurance. Other historic areas owe their character to a successful blend of ancient and modern, sometimes perhaps through facadism, where modern interiors are inserted to cater for those who require that type of accommodation.

CHARACTER, SENSE OF PLACE AND MEANING

Urban conservation is a process involving, first an appreciation, and then the creative protection of the particular character of parts of the built environment that have special architectural and historic significance. An appreciation of character involves a consideration of many factors including age, history, scale, style, form, texture, colour, uses, townscape grain, uniqueness and atmosphere. One of the most significant elements of character is the sense of place, which may be perceived particularly through design, form and colour, the spaces and linkages between buildings and the uses of buildings,[18] all of which combine to produce a distinctive spatial and visual quality, helping to distinguish one place from another and create local identity. A sense of place conveys meaning to the individual about his or her position in the environment, and allows identification with a particular space and the sense of drama, peace or whatever response is generated. In considering the factors which contribute to the character of an area, it may be argued that the retention, rebuilding or replication of particular facades may play an important role in the conservation process.

Age and history in themselves bestow a certain character on urban areas and its outward expression through the facades of buildings is a vital quality, the preservation of which is a fundamental part of the conservation process. Of course, it must be recognised that the older the built fabric the more difficult it becomes to justify preserving merely the exterior of buildings, because the age and possible rarity of the interior fabric is likely to make it equally valuable. Facades also play an important role in conveying historical associations, and there may be occasions where the retention of just a facade, as a reminder, say, of a famous person or an example of a particular architectural or design style, may be justifiable. This might apply where an association relates to an otherwise mundane building and the retention of the whole building cannot be withstood economically.

The design and form of the facades of buildings is an important element of character. Take, as examples, the order and harmony displayed in the architectural scenery created by the facades of the Wood family's Royal Crescent, the Circus and Queen Square in the city of Bath and the terrace architecture of the Adam brothers, Nash and Decimus Burton in London. The concepts of architectural scenery and terrace architecture have a time-honoured tradition in Britain and Europe. In seeking to preserve the character of townscapes where the provision of scenic settings for life's activities has been the fundamental role of the buildings, the most important objective must be the retention of the facades which create the backcloth. This perspective balances the criticism often made that facadism trivialises or undermines the process of urban conservation by treating townscapes as mere frontages, leading to a 'Disneyland' mentality.

Disneyland belongs to the world of entertainment and fantasy. It is superficial and does not deal in realities. The criticism that facadism turns townscapes into Disneyland seems to imply either that facadism results in buildings which are purely superficial, or that there is no place within the historic built environment for buildings to entertain and create fantasies at the superficial level. Dealing with the first point, this denies the fact that the main role of many historic buildings has been as architectural scenery in the theatre of life. Also it discounts the ability of creative designers to produce new spaces behind the historic backdrop which are capable of enriching the quality of people's lives. Turning to the second point, it should be recognised that many historic places reflect the rich diversity of the tapestry of life and there is no reason why, in appropriate circumstances, characteristics such as humour, fantasy, surprise and innovation should be denied. Not every building need be what it seems from the outside. Indeed, the outward display of style on the facades of many historic buildings is nothing more than frippery, in that behind the richly ornamental or finely crafted facades lie mundane, utilitarian interiors.

In the county of Wiltshire, the National Trust owns a village called Lacock where strict controls over alterations to buildings are imposed to preserve the character of the place. It is a functioning settlement, but the atmosphere there is of a place where time has stood still. A visit to Lacock is such an unusual experience that a sense of unreality pervades the place. The buildings create the impression that one could be walking through the film set for a historical drama. Lacock is a beautiful and remarkable example of a village which superficially has been preserved intact, and, as a place of historic interest, deserves the protection it receives. It is, no doubt, enjoyed by its inhabitants and many of its visitors. Inevitably, behind the facades of the buildings of Lacock there are the trappings of modern life. The outer visual impression is not the inner reality. While the exterior fabric has preserved the familiar local scenes, inside the buildings people live in the style of the twentieth century. There is indeed a sense of facadism about the place.

I referred above to the ordered harmony of the Wood family's work in Bath. In comparison, the character of King Street in Bristol derives from the rich diversity of the facades. Here, a visual riot of differing styles, colours, textures, detailing and form creates a streetscape of enormous richness and interest and a unique atmosphere (see Plate Section, xvi). In the same city, we return to an ordered setting at Brunswick Square – where the southern frontage was preserved as part of an office redevelopment scheme; here, the facades give the composition balance and formality. Remaining with the focus on Bristol, let us briefly consider the case of two Georgian buildings gutted by fire in Portland Square. All that remained were the facades of the buildings. This posed the question: would the character of the conservation area within which the formal square stands be best preserved and enhanced by either demolishing the facades and constructing a contemporary building or retaining the facades and erecting new accommodation behind? Bristol City Council Conservation Officer, Alistair Brook, commented in 1986 that to 'destroy a formal composition for the whim of change is surely "unreasonable" and most of the Public would be with us'.[19] Facadism won the day on this occasion, and the argument that such a course of action has contributed to retaining the original unity of the square is convincing.

Detailing and ornamentation on facades also lend character (see Plate 11). Familiar classical features such as cornices, columns and pilasters give order, beauty and interest. Mouldings articulate the lines on a facade, giving enrichment and meaning to the design. Shadows cast by features add further aesthetic dimensions, creating contrasts of light and shade, adding shapes and emphasising the textural qualities of the facade. The scale of facades is also an important characteristic. The monumentality of the facade of Atrium Court, fronting West Campbell Street, Waterloo Street and Wellington Street in Glasgow, evokes the sense of power and wealth of the Victorian city (see Plate Section, i). In comparison, the facade shown in the photograph of George Street, Edinburgh (see Plate Section, viii), is distinctly domestic in scale. Both these facades were preserved when redevelopment took place behind.

Britain, being one of the most geologically diverse countries in the world, has a rich variety of natural building resources. Traditionally, buildings were constructed in materials which were easily obtained in the area. These materials possess unique hues,

Plate 11 Detailing and ornamentation on facades such as this Victorian example of the Bristol Byzantine style at Victoria Street, Bristol, lends beauty, interest and character to the townscape.

tones, textures and weathering qualities which contribute to a sense of place and local identity so important to the character of the historic environment in the UK. Consider the importance of the hewn granite facades to the character of Truro in Cornwall. The Georgian glories of Bath were created out of the mellow, pale golden, smooth-faced, local oolitic limestone. The rugged austerity of Carlisle derives from the deep ruddy complexion of the stone buildings. Chester's ancient High Street is dominated by timber and plaster facades. The standardisation of building materials in the twentieth century and the distribution of concrete, machine-made bricks, reconstituted stone and claddings of plastic and metal to all parts of the country has eroded the feeling of local identity and created a mediocre 'anytown' syndrome.

In addressing the important contribution made by traditional facades to the character of townscapes and the arguments in favour of retaining them, the willingness, ability and opportunity of architects and clients in the present age to produce buildings displaying the architectural quality, identity or embellishment of buildings of the past must be considered. The realities of the contemporary development process in Britain suggest that only in exceptional circumstances will such quality be achievable in the foreseeable future. Building craftsmen still exist, although far fewer in number than before the modern era. Also machine technology enables the mass-production of traditionally styled embellishments for the exterior of buildings. However, in comparison with the use of contemporary materials and methods, the costs of natural materials and craftsmanship are generally prohibitive. In response to this situation there emerges an argument in favour of retaining features which are valuable to the character of an area and would be costly, difficult or impossible to recreate, such as important facades. Also, where it is not feasible to totally rebuild copies of derelict historic buildings within sensitive townscapes on the grounds of cost, it may prove possible to finance the construction of either a replica facade or one which evokes an appropriate style as a frontage to a building, the rest of which utilises less expensive contemporary materials and methods.

There is a danger that drastic change to the cellular nature of urban form and the 'grain' of the townscape, and also the consequential removal of existing uses, could, in many circumstances, inflict considerable devastation on the character of urban areas. Townscape 'grain' is dictated by plot sizes and shapes and frontage widths. A characteristic of some forms of facadism is that this 'grain' is significantly altered. The cellular nature of townscape is a vital part of its character and in the context of urban conservation, facadism in certain cases may be seen to be undermining objectives. In the case of Nos 1–6 Brunswick Square, Bristol, the tight or narrow grain formed by the terrace of Georgian houses has not been reflected by the development scheme, which retained the facade in front of modern open-plan offices. Instead of the traditional rhythm of individual houses indicated by the party quoins on the facade, one now sees, on close inspection from the square, the unbroken expanse of office space lit up by the ubiquitous fluorescent strip-lighting.

The Royal Fine Art Commission has highlighted the problem of the relationship of

Plate 12 John Outram's pumping station, Isle of Dogs, London.

the facade to the spaces behind not being recreated in cases of facadism because of the type of accommodation required by developers, and has suggested that a more appropriate approach to conservation would be the retention of at least one room-depth behind the facade.[20] This method has been used, for instance, in a number of cases in George Street, Edinburgh. The thicker veneer proves more convincing in keeping the character of the street intact than simply retaining just a facade. At least from the street frontage, the former townscape 'grain' is more authentically reflected. However, to take the analogy with wood a step further, just as the grain in a piece of wood is not simply on the surface but of the whole, townscape 'grain' is composed of whole building plots and not merely frontages. Therefore, unless redevelopment schemes repeat the complete cellular pattern of historic development, the characteristic townscape 'grain' will inevitably be eroded. To fully experience the character of townscape, it is often necessary to wander off the well-beaten tracks and main thoroughfares, to stray into back alleys, to obtain views of rear elevations and over roofscapes. Only then will the 'grain' of townscape be fully exposed and the impact of approaches to urban remodelling such as facadism understood.

The facades of certain buildings play an important part in the uniqueness of townscapes. Through their visual qualities, these facades remain in the mind of the observer and act as a focus for recalling townscape images from the memory. All sorts of qualities make facades special. Consider the beautiful ornamental patterns, shapes and rhythms on the front of the Church of Sta Maria Novella in Florence or John Outram's postmodern play with what appears to be Egyptian-influenced symbolism on the face of his Isle of Dogs pumping station in London (see Plate 12). In Helsinki, the monumental figures at the station entrance etch the facade in the memory. The facade of Wood's Royal Crescent in Bath has been mentioned several times elsewhere in this book, but its uniqueness is especially relevant to our thoughts at this point. The harmony both of its elegant design and its relationship with the landscape are truly special.

Urban environments having a distinct character and sense of place are more meaningful to the people who experience them. The significance of meaning in urban design and the relationship of the built environment to human behaviour was highlighted by Amos Rapoport in *The Meaning of the Built Environment*:

> when the lay public complains that churches, post offices, banks and so on no longer look like churches, post offices or whatever, one of the things they are saying is that the expected behaviours are not clear, and also that the designers have neglected meaning.
>
> (Rapoport 1982)

One important dimension of the urban conservation process is the aim of keeping intact townscapes that people understand and can relate to, where the cues and meanings are clear. To have meaning, townscape must be recognisable. Within the powerful emergence of the urban conservation movement in its early days, and the popular reaction against the featureless concrete urban landscapes created under the banner of International Modernism and planned comprehensive redevelopment schemes, was a desire to retain familiar street scenes with their own identity displayed in a diversity of elements such as style, colour, form, craftsmanship and ornamentation. The intentionally anonymous modern buildings often lacked recognisable facades. Characterless townscapes were created, bereft of interest and a sense of place. Buildings and places lacked meaning, thus denying people an opportunity of relating to the local identity of their built environment.

Facades are highly significant elements within townscapes in terms of people's experiences of the built environment. Familiar scenes, vistas, landmarks and symbols communicate cues and generate a diverse range of feelings. Consider the sense of theatre created by the majestic sweep of the facade of Bath's Royal Crescent, the awesome scale of the facades in Haussmann's Paris boulevards, the excitement of the imposing facades which form the lovely vistas in central Edinburgh, the feeling of tranquillity in the dignified order of Georgian streetscape in many British towns and cities, and the inspiration given by a beautifully crafted historic facade. Even the facades of the more humble historic buildings can perform valuable functions by conveying meanings, forming local identity and generating feelings that are precious to many people. The unpretentious stone fronts of Sir Titus Salt's dwellings at 'Saltaire' in Yorkshire convey a feeling of solid stability and their historical associations remind us of the founder's philanthropic attitudes. Old industrial buildings, for instance, which may have accommodated businesses that provided an important source of employment and an economic basis for local communities can convey a sense of continuity and reassurance during uncertain times of change, even though they have been adapted or converted to other uses. An elegant Edwardian shop frontage often evokes a sense of style and dignity missing from modern counterparts and such features can be accommodated successfully in new shopping developments or converted to a range of alternative uses. Clearly, in certain circumstances where particular buildings are under threat, for example, because of structural decay or damage, or because they are economically redundant, the preservation or rebuilding of the facades, or possibly the exterior shell in the case of freestanding buildings, will be of fundamental importance in protecting the special character and sense of place of historic townscapes and the often unique cues and meanings given out.

In some cases, the facades of former buildings have been preserved almost as architectural sculpture or monuments to the past. At Liverpool's Albert Dock there is the interesting case of the entrance of a former tram shed being preserved as a freestanding granite arch. An imposing feature in its own right, it adds visual interest and conveys meaningful messages of the dock's former use. Similarly, at Marples Wharf, Bath, several arches, which were once part of the facade of riverside industrial or warehouse premises, are retained as a feature. The arches perform an enhancing role in preventing a gap site breaking the continuity of fabric, much of which is rich in character, along the water's edge and distract the eye from some of the more mundane Bath townscape which would otherwise be more rudely exposed (see Plate Section, xviii). Jerry DeMuth, writing in the *Estates Gazette*, refers to the 'sculpture approach to facade preservation' in relation to a number of examples.[21] In Salt Lake City, Utah, in 1971, a proposal to demolish a department store dating from 1876 apparently caused a statewide protest. This led to the restoration of the three-storey cast-iron facade, not as an integral part of the replacement store building but as a feature stuck onto a plain concrete wall. In Philadelphia, an Egyptian Revival-style marble facade from around 1902 was demolished and, three years later, rebuilt as a feature over which the 21-storey Penn Mutual building cantilevers. A parking lot was hidden behind a stone and brick facade rescued from the remains of a fire-damaged building in Portland, Oregon. There is no reason why, in terms of the spirit of conservation, attractive features of buildings of the past should not be used to enhance present-day development. Their useful preservation is consistent with conservation principles. However, these features cannot always adequately contextualise new development where, for instance, it is out of scale with the traditional townscape or destroys its characteristic grain. In such cases, both old and new fabric appear ill-at-ease, and there is the danger that the preserved feature becomes trivialised and an object of ridicule.

REFUGE IN THE PAST

An important aspect of urban conservation is the perceived relationship between old and familiar townscapes and human wellbeing. In this relationship, long-established urban fabric provides identity, stability and continuity, particularly in times of uncertainty or rapid change. In England, central government recognises that 'Public opinion is now overwhelmingly in favour of conserving and enhancing the familiar and cherished local scene' and urges local authorities to take this into account when framing policies (Circular 8/87). P. F. Smith considers that 'Familiarity breeds contentment'[22] and Worskett links the retention of architectural and historic features with the maintenance of local identity and the importance of this to the quality of life.[23] Cantell points to a psychological need for the continued existence of familiar surroundings and the tangible assurance of stability given by bricks and mortar.[24] The symbolic element of townscape is also important to the quality of life (Rapoport 1982). Wherever buildings form the background to experiences they assume meanings related to those experiences and become permanent symbols of transient events. However profound the events, memories fade with time and so the removal of the buildings affects the recall of experiences. Also, as Smith argues, people have a need to test from time to time their roots in the past,[25] and Lord Clark states that civilised man 'must feel that he belongs somewhere in space and time: that he constantly looks forward and looks back'.[26] Old buildings help to consolidate an important awareness that people acquire of the evolution of society. Historic towns displaying the slow growth of centuries are expressive not only of the link between the past and present but also of the deep roots of our society.[27]

In the light of the above ideas, it can be seen that the psychological effects of drastic and rapid changes to historic built environments are considerable and inevitably undermine the quality of many people's lives. The realities of a mixed economy in the contemporary world often mean that the pressures for change in historic areas are irresistible. Only by having land, capital and the means of production in the hands of a controlling and co-ordinating centralised agency – such as the state, to use a now unfashionable term – could a process of urban conservation which preserves whole historic areas completely intact, including their local economies, operate effectively. Without such control, urban conservation must therefore be concerned with preserving the essence of an area's character and retaining sufficient historic fabric to fulfil people's psychological needs and uphold the quality of their lives. Historic townscapes provide an effective therapy for the culture shock created by the rapid pace of change that occurs in the late twentieth century. Facadism is an important concept here, because it can enable people to continually enjoy and relate to familiar street settings, even though it has been necessary to redevelop behind the frontage. The facades of old buildings create identity, act as symbols and landmarks, provide a sense of continuity and stability within townscapes and may still enable the recall of experiences long past, even though the rest of the original fabric has since disappeared.

Facadism presents an opportunity for the successful union of old and new architecture. Through such a union, people are able to see their place in time, with images of the past providing reassurance and the new fabric reinforcing the relevance and value of the present age.

Having recognised that facadism has a role to play in providing vital links with the past, great care is needed to preserve sufficient historic fabric to show clearly how societies have evolved and to provide people with a meaningful context for their place in time. There is a need, for example, to retain fabric displaying ancient street patterns, plot ratios and the interior plan form of buildings. There is also a need to protect complete examples of representative types and styles from different eras both in terms of individual buildings and areas of townscape. The above considerations were very much part of Bristol City Council's decision during the 1980s and early 1990s to resist facadism in relation to a terrace of properties in Victoria Street (Nos 10–22) where, during the nineteenth century, the attractive facades were added to what is, in terms of Bristol, a rare survival of medieval development (see Plate Section, xvii). Although the facades are worthy of protection in their own right for their visual interest and architectural value, the loss of the fabric behind would sadly erase important contextualising evidence of the area's medieval past.

It is often said in support of conservation that historic townscapes display far greater aesthetic and artistic qualities and more human scale than contemporary development. Such an argument displays a lack of faith in the architecture and urban design of the present and is symptomatic of an age of uncertainty and insecurity. Dobby, in *Conservation and Planning* (1978), succinctly captures these numerous motivations behind the conservation movement when he refers to 'yearnings for symbols of permanence and security from a perhaps artistically finer past'. The 1950s and 1960s represented a time of relative optimism in many walks of life. The 'permissive society' was challenging old-established order. New ideas in music, painting and fashion were indicative of the mood of the time. In the built environment, comprehensively planned redevelopment schemes and Modern movements in architecture were continuing to change the character and appearance of urban areas throughout Europe. By the 1970s, however, British society was beginning to experience an increasing feeling of demoralisation. There had already been the devaluation of the currency in 1967 and the atmosphere of decline was emphasised by the major economic hammer blows of the oil crisis in 1973 and Britain's submission to the dictates of the IMF in 1976. The monetarist economic policies of Thatcherism during the 1980s failed to avert a recession, which by 1992 was still continuing to deepen and was further undermining the nation's self-esteem. Instead of modernisation during the last twenty years, the overall perceived trend has been towards rationalisation, redundancies and de-industrialisation.

Widespread dereliction has been a major symptom to affect parts of the historic built environment. The perception in Britain of a failure to deal with a range of problems since the war has created an atmosphere of social and economic decline, which has fuelled support for the process of urban conservation as people, lacking confidence in the present or future, seek refuge and inspiration in symbols of the past. The preservation of what remains of familiar historic townscapes therefore provides solace for a disillusioned and disoriented population. Robert Hewison in *The Heritage Industry* (1987) talks of Britain being reduced to manufacturing 'heritage' in an attempt to dispel the gloom created by the deep social convulsions and economic decline since 1945. Our past culture has become an economic resource to be exploited in a ready market created by all sorts of insecurities and uncertainties about the present and future.

In some respects the conservation process and the heritage industry are comfortable and mutually supportive partners. Obsolete historic buildings often provide ideal accommodation for museums, exhibitions of memorabilia, industrial heritage centres, theme attractions and so on. In return, these heritage uses fund the retention of the buildings. It seems ironic, however, that many of these buildings, which once contributed to the nation's economic growth and created employment opportunities through the production of actual commodities, are now places which tend to perpetuate a spiral of lost confidence in both the present and the future rather than encouraging economic and social recovery. Many of the people who visit these places are likely to be disillusioned by Britain's industrial decline and the other economic problems of the present era. They relive what is often an idealised version of the past with the unsavoury or unattractive elements filtered out; consequently, the past may well become much more attractive than the present. It seems ironic that in many preserved historic buildings where people were once employed manufacturing tangible products, people, many of whom are likely to be or have been unemployed, are now sold nostalgic experiences and the inevitable souvenirs of their past. There is a danger that Britain's future cultural vitality will be sorely debilitated by this preoccupation with the past, in that the climate is not conducive to the emergence of new ideas and a growth of confidence in the present and future. Conservation is a dynamic process which should embrace new cultural development. There is a sense in which a vast explosion of the heritage industry can be seen as a stultifying influence.

It is interesting, in the context of this book, to consider whether facadism is part of the mentality which permeates much of the heritage industry. Certainly, facadism has been a symptom of a lack of confidence that contemporary architecture could produce streetscapes of an equivalent visual quality to those that would be replaced. There is considerable justification for this attitude when we compare the craftsmanship, beauty and character of the facades of many historic buildings with much post-war development. Nevertheless, by preserving merely facades – sometimes the most attractive and architecturally valuable features of buildings – or images of past styles in facsimile facades, we need to ask whether or not the past and the process of conservation are being trivialised because only the superficial elements of historic townscape character are being represented. After all, urban character is composed of a combination of diverse elements, some outwardly demonstrative, others more discreet and less obvious. Some will be obviously attractive, others less appealing. However, all these aspects are part of the synthesis that creates character.

It could be argued that facadism is a process of filtering out the less spectacular parts of past development which creates a distorted image of historic townscape. There is also a danger that facadism encourages the idea that conservation is a process of constantly prettifying or gentrifying places, and a form of visual disinfection. In the same way that the heritage industry often makes the past appear more desirable than it actually was, it could be argued that facadism is a concept that encourages the glorification of buildings of the past by simply keeping the impressive street frontages and removing the more mundane fabric which not uncommonly may be found behind the facades. It is often the 'Bohemian' nature of places which give them their character. We must regularly look beyond street frontages in assessing townscape quality and character. On occasion we may find that the chaotic jumble of fabric behind facades is in fact highly stimulating in a visual sense – perhaps because of the diversity of shapes, colours or textures – or historically valuable because of its age and rarity. The fabric may also be important because it provides affordable accommodation to the users which help to create the characteristic atmosphere of the locality and sustain its local economy.

Preserving facades and redeveloping behind, of course, provides an opportunity to tidy the townscape, which some people may argue is a form of enhancement. However, so much that may be vital to the character of a historic area could be lost by such action. While, for instance, history may be recorded before old fabric is removed, there is no substitute for the tangible experience of seeing and touching the work of architects and builders of past eras and standing amid what represents our roots within the built environment. Obviously, there will always be occasions when historic buildings suffer from problems of severe disrepair, but the advances in building technology which allow old fabric to be effectively treated, strengthened and restored mean that resorting to demolition, or possibly facadism, is not necessarily essential.

It could also be said that too much facadism is a hindrance to cultural development in that it denies contemporary architecture an opportunity to evolve and fully express itself. If new buildings are continually hidden from view behind facades of the past, and thereby visual images and statements of the present or the future excluded from historic streetscapes, the onlooker is denied an opportunity of discovering that buildings of the present and future have architectural validity. There is a need for courage on the part of decision-makers to allow new styles to infiltrate traditional urban settings. This after all is a process that has continued ever since the first urban settlements. I referred earlier in the book to my visit to Edinburgh in the spring of 1992, and the development at St Andrew Square, where Scottish Provident Institution, the site owners, were extending their accommodation behind the preserved facades of Nos 4 and 5. Roland Wedgwood's exciting contemporary statement, apparently based on a Belgian priest's theoretical concept of aesthetic proportions, would, I believe, have achieved a great deal in terms of demonstrating the value of new architecture within the historic urban setting. The design displayed deference and sensitivity in terms of its setting, but also clearly stated its place in the late twentieth century.

While many people undoubtedly support the notion of urban conservation, they also desire the comforts and conveniences of modern amenities. In order to achieve such amenities it is sometimes necessary to significantly adapt historic fabric, which encourages developers, architects and town planners to look closely at facadism as a solution to urban conservation problems. The traditional outer skin provides continuity and reassurance to people on the street, while behind the facade the replacement of existing fabric with new will, in some cases, make the provision of the required comforts and advantages of modern amenities more economically viable.

Although it is understandable that the crushing mediocrity of vast numbers of Modern and postmodern buildings has led to communities seeking refuge in the past, this perspective has to be balanced against the fact that contemporary development can succeed in enhancing historic townscapes. A visit to Rouen in 1987 with a group of fellow students reinforced a personal faith in the value of good modern architecture and the possibility of successfully integrating it into historic environments. It also demonstrated to me how important it is to have the courage to allow the architectural continuum to develop in the historic environment, otherwise the chance to create experiences such as that in Rouen will be missed. In a town square in the heart of the lovely medieval core of Rouen stands a modern church, which, when first seen provides a violent yet uplifting culture shock. After exploring the historic fabric of the city for a number of hours, the refreshment of finding a modern building of such beauty was most satisfying. This was a view expressed by a number of my companions. The character of the square in Rouen has not been destroyed but rather enhanced through adding a new architectural dimension (see Plate 13).

Plate 13 Modernist church in the medieval core of Rouen, Normandy.

Two different aspects of the facadism–conservation relationship flow out of the Rouen experience. First, it demonstrates the principle that in order to conserve historic settings we need not necessarily ape or evoke the past in new development or shield the new from view behind a traditional facade. Second, conservation can, in principle, successfully involve old and new in juxtaposition. This applies in cases where new buildings are completely separate entities, as in Rouen, or where old and new fabric are joined or styles mixed in the creation of a new building as, of course, is the case where facadism occurs. Generalisations are impossible about the most appropriate way forward in any particular situation. What we should not say is that we can never do this or do that, but rather an openness of mind should always prevail, and an awareness of the range of possibilities and options available. Sometimes facadism will be right and on other occasions totally inappropriate.

The Alec French Partnership achieved a tolerable union of Pope's facade with imaginative modern office accommodation at Brunel House in Bristol. Here, the front and sides of the building are clearly visible from the street and there is no question that old and new fabric are separate entities. The most disappointing aspect of this building is its monolithic roofscape, which fails to reflect the scale and interest of its predecessor. Sadly, roof lines are often radically altered or destroyed as a result of facadism, a point that should not be overlooked in considering the impact of new development on townscape character (see Plate Section, vi, vii and xix).

As far as the public is concerned, Marks and Spencer achieved a good contextual solution to building their city centre store in Exeter during the early 1980s. In comparison, however, the *Architects' Journal* condemned the scheme as a 'dog's dinner'. The fact that the development was so well received by the public makes the outcome a successful conservation exercise, or so it may be argued. A typical response from a person in the street in Exeter was 'I like the building very much. It fits in splendidly with Exeter'. Many people interviewed liked the detailing. The scheme relied to a considerable degree on the past (and facadism) to achieve its appeal, breaking the large external envelope into a series of smaller units, some of which were reconstructed in traditional style by using modern techniques. A listed lath and plaster facade (Queen Street), which had deteriorated to a dangerous condition, was rebuilt in rendered brickwork and pre-cast concrete sills, mullions and window heads. A stone loggia building (the Woolmarket building and Archway) was replaced by a reconstituted stone version. Alongside these traditional elements are more contemporary facades, most of which are inoffensive if not inspiring. It was possibly a measure of the public's low expectations and lack of confidence in terms of the visual quality of new development, rather than the architectural merit of the buildings, that the people interviewed in Exeter found the Marks and Spencer scheme so appealing. However, the example illustrates what is in my experience the case, that developments reflecting the past through the retention or rebuilding of traditional features and ornamental details will win the widespread support of the public. This begs the question whether or not such support therefore justifies facadism as a valid form of conservation. The answer must be in the affirmative if such an approach brings pleasure to many or fulfils a need in people for security, reassurance and a refuge in the past. Having acknowledged this, it is important for those who make recommendations and take decisions affecting the historic built environment to consider whether facadism is enough to truly conserve a particular environment and its special qualities. There will be occasions when it would be too superficial an approach, either because insufficient fabric is being retained or rebuilt, or the form of a new development does not accurately capture the character of surrounding townscape, or because facadism is denying an opportunity to create a wholly new but enhancing design form.

Two recent developments in the city of Bath provide contrasting approaches to urban conservation. One involves facadism, at 'The Colonnades', where a large covered modern precinct has been created behind a retained classical street frontage. In the case of the Seven Dials scheme, a modern building was demolished and replaced by completely new development in a design, scale and form to reflect the surrounding historic townscape. Both schemes preserve and enhance the character of Bath in different ways. The former contrasts the traditional elegance of the classical colonnade along the street frontage with a stylish late twentieth-century interior, whereas the latter faithfully recreates a historic development pattern.

ENHANCEMENT

Enhancement has become a key word in the 'official' view of conservation. Inevitably, views on what constitutes enhancement in terms of the character of areas differ widely. From one viewpoint, enhancement could simply mean improving the appearance of townscapes through the removal of eyesores or tidying and cleaning existing fabric. From another, it may involve fundamental physical, social and economic change. Whether or not facadism has a part to play rests on one's perspective of enhancement. It could be argued that, if the objective is to bring tidiness to townscapes, facadism may on occasions have an important role in enhancement. It is not uncommon to find behind an impressive facade an architecturally mundane jumble of built fabric which many people would refer to as eyesores. Facadism provides the opportunity to retain impressive facades and to create a more ordered form of development behind.

In terms of helping to remove dereliction from historic townscapes, facadism has been an important concept since the early 1970s. For example, in some situations developers have been unwilling to invest in accurately reconstructing historic buildings, but have agreed to make some gesture towards protecting the character of townscapes by financing the retention or rebuilding of facades or the outer shells of former buildings as part of redevelopment proposals. It is likely that in many instances facadism has provided the key not only to unlocking the development potential of a derelict site but to saving whatever was realistically salvageable of the historic fabric and in this respect has helped to further the cause of urban conservation (see, for example, 'Case Study 1 – Brunswick Square' in Chapter 7).

A number of interesting and relevant issues are raised in the case of Bathurst Basin, Bristol, a former industrial dock which has been transformed into a quayside residential area and marina. Critics of the process of change which has overtaken the area lament the demise of its industrial character. Others express the view that the basin has been opened up to the public, its visual character greatly enhanced and its industrial tradition reflected in the preservation of the facades of two former warehouses (behind which a squash club has been constructed). Depending on the perspective taken, facadism, in this case, can be seen as either (i) a useful conservation technique of retaining a townscape element which has its own intrinsic value (see Plates 6 and 7, pp. 50 and 52) and one which also serves to remind present and future generations of the industrial past of Bathurst Basin, or (ii) as part of the regrettable process of changing an industrial and employment resource into housing and leisure space for the middle classes. It should be noted that the basin already contained some modest nineteenth-century terraced working-class houses and so, leaving aside the issue of occupier group, the introduction of housing was not necessarily an alien principle. As with the rest of Bristol's City Docks the industrial uses had severely declined and, prior to redevelopment, the air of dereliction hung heavily over Bathurst Basin. Perhaps in a fully planned economic system, through public investment, an employment function could have been maintained in Bathurst Basin and any new housing allocated to a different occupier group. It might also have been possible to retain the Byzantine-style warehouses intact to accommodate a use which more accurately reflected the building's past than a squash club. However, accepting the realities of the market-led system and the investment decisions within the development process, planning authorities, amenity groups and the public have to come to terms with compromise solutions. The outcome at Bathurst Basin has some merit as an urban design and conservation exercise. The Byzantine facade, with its rhythmical beauty, dominates the scene and its retention has achieved a most important conservation objective. It is knitted relatively well into the subservient adjacent housing, the form and massing of which is in keeping with the modest nineteenth-century dwellings in the basin and Bristol's maritime position. The most disappointing design aspect is the relationship to the facade of the new squash club building, which appears out of scale with its impressive facade.

BEHIND THE FACADE

Clearly, another important consideration is that built fabric behind the facades of many listed buildings is of equal significance architecturally and historically to exterior fabric. In some cases it is of greater significance. In such circumstances facadism would be an inappropriate approach to conservation. The view of English Heritage – that it is the whole building which is normally listed, as opposed to separate features, such as a facade – was referred to earlier, and this view emphasises the need to always look beyond the outward form and appearance of buildings in assessing their heritage value. Internal features such as fine plasterwork, timber panelling, ornate fireplaces and ancient forms of construction may lie behind facades which, although interesting in their own right, are less valuable historically or architecturally. The plan form of old buildings provides historical evidence of how buildings have evolved and, in conservation terms, to preserve such information should be of fundamental concern. The internal fabric of buildings does not make the public impact of exteriors, but should receive appropriate consideration and protection.

On Victoria Street, in Bristol, between the city centre and the main railway station are a number of Victorian facades in the Bristol Byzantine style. To many people the facades would appear far more interesting and valuable than the dilapidated jumble of fabric to the rear. In reality, the facades hide a rare example of medieval burgage plots, the loss of which would be extremely regrettable in terms of the historical significance of the buildings. In 1987, the Curator in Field Archaeology at the Bristol City Museum and Art Gallery expressed the view that the fabric behind the facades represents 'a rare survival of the structure and flavour of the medieval town'.[28] Sadly, most of the buildings have been empty for some time, a sign that the former small-scale retail and commercial uses found it difficult to remain viable in this location, which is isolated from both the main shopping centre and commercial heart of the city. Consequently, the structural condition of the buildings has continued to deteriorate (see Plate Section, xvii). Developer pressure during the 1980s to remove everything apart from some of the Victorian facades was resisted by the local planning authority and a local amenity group produced a convincing viable refurbishment scheme, retaining the medieval fabric, to demonstrate that facadism was not necessary. Discussions with the developers in 1988 indicated that, in their view, facadism was the only viable redevelopment option. Repairing the original fabric behind the facade was considered too costly, given that their marketing research had shown a demand for larger office spaces in the locality than could be achieved by restoring the old small-scale buildings. However, to date, the site remains close to dereliction.

The Mappin and Webb site, London

One of the most significant recent cases in planning law to embrace a number of fundamental urban conservation considerations relates to the Mappin and Webb site, adjoining the Mansion House and Bank of England, in the heart of the City of London. Also, in the context of this book, the case is of relevance to an examination of the role of facadism in urban conservation. A particularly notable feature of the case is that, in relation to one of the City of London's most highly sensitive areas of townscape, the Secretary of State for the Environment departed from normal 'official' British urban conservation policy – within which there is a strong presumption against demolition of listed buildings, particularly where every effort has not been made to continue the existing use or find alternative uses – and allowed an appeal by the site owners, following a public inquiry in 1989, for the demolition of eight listed buildings and other buildings in a conservation area and the construction of a new building by British architect James Stirling.

The extremely prominent site, which lies at the Bank of England intersection where seven major streets meet, is roughly triangular in shape, the northern boundary being Cheapside and Poultry, and Queen Victoria Street forming the south-eastern limit. Several narrow lanes pass through the site, which is within the designated Bank conservation area. The eight listed buildings standing on the site have Grade II status. Circular 8/87 advises that such buildings are 'of special interest, which warrant every effort being made to preserve them'. The group value of these particular buildings was also identified in the listing. The Bank conservation area also contains five Wren churches and Hawksmoor's St Mary Woolnoth.

The owners of the site, Number 1 Poultry Ltd and City Acre Property Investment Trust Ltd, had been seeking planning permission to develop the site since 1981. Their view was that the site required an architecturally distinguished modern building which would enhance the character of the area. The view of interests claiming a conservationist perspective was that the only acceptable approach would be a scheme of internal reconstruction and refurbishment to preserve all those external architectural features of the listed buildings that had resulted in the buildings being listed. In 1981 the owners announced a scheme, which also included No. 1 Queen Victoria Street, to demolish eight listed buildings and other unlisted buildings, erect a 290ft tower block designed by Mies van der Rohe and create a square in front of the Mansion House. The development proposals were rejected both by the local planning authority and, following a public inquiry in 1984, the Secretary of State for the Environment. In his written decision, the Secretary of State rejected the Mies van der Rohe proposals on the grounds that it would 'dominate the appeal site and the surrounding area to a wholly unacceptable extent because of its height and bulk and because of the stark conflict between it and the scale and character of neighbouring buildings'. He did not, however, 'rule out redevelopment of this site if there were acceptable proposals for replacing the existing buildings'. Also the view was expressed in the decision that 'it would be wrong to freeze the character of the City of London for all time'.

In 1988 the owners of the site returned with the scheme designed by James Stirling. This was also rejected by the planning authority, but subsequently allowed, following another public inquiry by the Secretary of State, who supported the appointed Inspector by agreeing to allow the demolition of eight listed buildings on the basis that the proposed new development would not only be appropriate visually and architecturally to the characteristics of the surrounding buildings but would also 'preserve the character of the area as a whole and could positively enhance it in some respects'. The new development was considered a 'worthy modern addition to the architectural fabric of this part of the City' and the view was expressed that 'it would contribute more, both to the immediate environment and to the architectural heritage, than would the retention of the existing buildings', which were regarded as being relatively less important than the surrounding Grade I listed buildings and Grade A churches.

A crucial factor in this case appears to have been the fact that the appointed Inspector considered that a great opportunity to build 'a considered mature work by a British architect of international stature of whose achievements the nation can be justly proud' in the core of the historic heart of the City would be lost if the appeal were dismissed. The Secretary of State accepted his Inspector's view, although stopped short of agreeing that the Stirling building 'might just be a masterpiece'.

It should be noted that the Secretary of State made it clear that the Mappin and Webb decision was not to be taken as a precedent to override the normally strong policy presumption against the demolition of listed buildings. However, the decision – which was challenged by Save Britain's Heritage (SAVE), a conservation pressure group, on a point of law, but finally won by the owners in the House of Lords in 1991 – demonstrates a principle that on occasions, albeit in special and exceptional circumstances, the process of urban conservation promoted by central government may involve the replacement of buildings of undoubted architectural and historic merit with new buildings.

The case also has another interesting conservation dimension: that is, the question of redundancy of historic fabric in an assessment of proposed development. SAVE attempted to demonstrate at the 1989 public inquiry that the useful economic life of the existing buildings could be viably extended by commissioning a scheme by Terry Farrell, which retained the facades of the existing buildings. The site owners accepted that the Farrell scheme was a viable proposition, but also stated that it was not their intention to follow such a course of action. Neither the appeal Inspector nor the Secretary of State chose to follow the advice in Circular 8/87 concerning the re-use of listed buildings (para. 89) which, had they done so, would surely have led to the appeal being dismissed. The Secretary of State's decision has drawn criticism from HRH The Prince of Wales, who in *A Vision of Britain* (1989) compares the Stirling building to 'an old 1930s wireless' and asks 'But why pull down one of the few remaining bits of the Victorian City, including no fewer than eight listed buildings, at all? What on earth is the point of having conservation areas if we are going to disregard them?'

In terms of the role of facadism in the process of urban conservation, the Mappin and Webb case raises the interesting point that, in some circumstances, even in the most sensitive of urban townscapes, conservation interest groups such as SAVE advocate facadism as a valid approach to urban remodelling, thus conferring a respectability on a concept which has, in relation to the achievement of urban conservation objectives, been considerably maligned. Clearly, in this case, the main contribution of the historic buildings under the threat of demolition is their townscape value as a group. SAVE's main objective appears to have been to preserve as much as possible of the existing external fabric, a great deal of which is rich in ornament, detail and craftsmanship. The Victorian landmark of J. and J. Belcher's Mappin and Webb building with its Franco-Flemish tower has enormous presence in the townscape and its replacement by James Stirling's monumental contemporary statement would undoubtedly have a significant impact on the character of the townscape. People experiencing the new townscape would receive very different messages compared with those conveyed prior to redevelopment, and the sense of place would be dramatically changed. By keeping the Mappin and Webb facades in place, as proposed in the Terry Farrell scheme, the traditional landmarks and inherent sense of place would remain, helping to familiarise people's experiences of the local environment. The new build element of the scheme behind the facades would breathe new life into the rest of the site and allow any necessary updating of the accommodation to reflect new social and economic demands, but with a sense of deference to the qualities of the existing townscape.

From a personal point of view, the above considerations leave little doubt that Farrell's facadism would reflect the spirit of urban conservation much more than Stirling's new building. So much which contributes to the character of the area would be lost by demolishing the existing buildings. This part of the city would be poorer without the visual interest imparted by their beautifully crafted exterior features, their contribution to the sense of place and the association of the buildings with the city's commercial past. Further, Terry Farrell's scheme shows that the provision of modern accommodation would be feasible without the loss of the familiar street scene. It is a shame that James Stirling's creative talent to design buildings worthy of important settings should result in the destruction of impressive and, no doubt, much loved traditional townscape when there are so many parts of Britain's inner urban areas screaming out for the touch of an artist.

COMPROMISE AND PRAGMATISM

The urban conservation process often requires compromise to resolve the inherent conflicts within historic built environments. On occasions the concept of facadism emerges as a pragmatic option which enables competing interests to achieve sufficient objectives and reach acceptable compromises. With this in mind, and also the architectural and townscape perspectives of facadism dealt with earlier in this book, to totally deny the validity of facadism as a form of urban conservation is to bury one's head in the sand not only to the realities of contemporary development processes but also the complex and contradictory nature of both architecture and urban design.

Plate xvi The colourful streetscape of King St, Bristol.

Plate xvii Medieval fabric which has been preserved behind Victorian facades, Bristol.

Plate xvii(a) Detail of the contrasts between Victorian and Medieval fabric.

Plate xviii Preserved arches at Maples Wharf, Bath.

Plate xix The former building on the site of Brunel House.

Plate xx Bond Street, Bristol – 1984 office development constructed behind the preserved facade of 1–6 Brunswick Square.

Plate xxi Vista at western end of George Street, Edinburgh, closed by the facade of a former church.

Plate xxii Former church in Charlotte Square, Edinburgh, at the western end of George Street, which has been completely remodelled internally to form a public records office.

COURAGE

7

WHY FACADISM?

Certainly since the early 1970s in the UK, it has generally been assumed that manifestations of facadism have been the result of decisions aimed at resolving the problems of integrating new development into historic environments or of preserving facades which have been considered to be valuable architectural features in their own right, while at the same time allowing urban areas to modernise. Facadism has, therefore, been seen in some circumstances as a means to an end in the achievement of urban conservation objectives. The previous chapter examined the validity of the concept of facadism in relation to the spirit and principles of urban conservation and considered circumstances when it may or may not be an appropriate approach to development involving the built heritage. In this chapter my intention is to explore the reasons behind facadism becoming a chosen development option, and to explain why, on some occasions, it becomes a reality.

We may speculate about the reasons for particular outcomes in different cases. For instance, does facadism sometimes occur because town planners promote it, either explicitly or implicitly, as part of urban conservation policies? Alternatively, have conservation policies been effective in resisting facadism in some cases? Is facadism the result of local authorities being negligent in monitoring the condition of their stocks of historic buildings? Does facadism occur because the implementers of urban remodelling see it as the key to unlocking the development potential of some historic buildings and sites and, therefore, does facadism indicate that economic considerations override good conservation practice? However, by detailed analysis of actual examples where facadism has been proposed and either implemented or, alternatively, resisted, we can go further than mere speculation and arrive at explanations of outcomes on the basis of empirically derived information. In the light of such explanations we are able, for instance, to reflect on the degree to which outcomes are either in harmony or discord with urban conservation policies.

The analysis of the subject probes deeper than simply assessing the end product of what is a complex decision-making process, involving interactions and negotiations between a range of actors and agencies including policy-makers and those upon whom the implementation of action depends. To fully understand the concept of facadism, and its contribution to the process of urban remodelling, it is necessary to consider the significance of certain events, actors and relevant circumstances that have led to it being proposed and, in some cases, implemented. Events may relate, for example, to particular meetings when strategies were formulated or decisions taken. The term actors is used to refer to key individuals or organisations who have influenced outcomes. Social and economic circumstances and the condition of particular buildings during decision-making processes may, for instance, be relevant factors in the outcomes in many cases. The approach taken in my research for this part of the book has been to adopt an 'action perspective', taking an implemented course of action as central to the analysis and seeking to establish how and why it occurred (Barrett and Fudge 1981). For the purposes of research, this implemented course of action in some cases referred to an example where facadism had occurred, in others it involved alternative action, or inaction.

Clearly, some outcomes within the development process appear to relate directly to expressly stated policies. Others apparently do not and are more of a reflection of situations encountered 'on the ground'. Within the process that leads to development action, public-sector town planners take a central role in policy formulation and decision-making. They are often seen in the role of passively allowing or actively promoting facadism as part of urban conservation strategies. However, to what extent do the views of local authority town planners reflect such a perspective? The following section summarises the results of a survey of planners working in thirty local authorities throughout England, Wales and Scotland in 1991, covering a range of urban area types

Table 7.1 Planners' opinions on facade preservation

Pro-facadism	
1	Where buildings have mediocre interiors.
2	Where facade is worthy of retention and rest of fabric is beyond repair.
3	If facade is the only original feature.
4	Where existing building cannot be used beneficially.
5	To protect townscape integrity, e.g. where large-scale interruption in street scene.
6	Acceptable because it maintains much of the existing visual character, architectural quality and form, while providing modern facilities behind.
7	Last resort compromise.
Anti-facadism	
1	Historical significance and integrity lost – buildings are three-dimensional.
2	Loss of valuable interiors.
3	Prevents modern/contemporary additions to street scene and continuity of architectural development.

Table 7.2 Planners' opinions on facade replication

Pro-facadism	
1	As an exception where continuity is important.
2	Preferable to loss of historic townscapes.
3	The public expects this approach.
4	Valid where buildings are part of a planned concept.
Anti-facadism	
1	Nothing authentic remains.
2	No replica facade is ever as good in quality, appearance and character as the original.
3	Difficult to comply with current building regulations and standards.

including historic gems such as Bath, Norwich and Edinburgh, large conurbations such as Liverpool, Manchester, Leeds, Glasgow, Swansea and Bristol, and small market centres such as Barnstaple (North Devon), Chippenham (Wiltshire) and Axminster (East Devon). A 50 per cent response to a questionnaire sent to the local authorities was received, and in some cases this was followed by discussions with the respondents. Of course, the opinions expressed are those of individuals and cannot necessarily be interpreted as official local authority views on facadism.

PLANNERS' ATTITUDES IN THE UK

It is interesting that none of the local authorities who responded has evolved any explicit policies relating to facadism, which is somewhat surprising considering that

Table 7.3 Planners' opinions on refacing buildings

Pro-facadism	
1	Can ensure retention of old fabric if facade is badly decayed.
2	New facade may improve appearance of twentieth-century buildings.
3	Environmental savings, e.g. externally unattractive buildings recycled rather than demolished.
Anti-facadism	
1	Makes nonsense of buildings as a whole.

Table 7.4 Planners' opinions on internal gutting and remodelling

Pro-conservation	
1	Where re-use is difficult other than by complete change to interior, and exterior shell is important to townscape.
2	Ensures saving of building materials.
3	Where interior has been greatly altered and no longer contributes to the character of historic buildings.
4	Protects integrity of townscape.
5	Same as facade preservation in principle.
Anti-facadism	
1	Excessive loss of interior fabric.
2	Inappropriate where valuable interiors are intact.

most of the planners expressed the view that facadism should be regarded as a 'last resort' alternative to the complete loss of valuable historic buildings.

Planners were asked whether or not they felt that facadism was generally well received by local communities and elected members in their areas, and approximately 50 per cent considered that facadism was acceptable in terms of popular opinion.

Four different types of facadism were presented to planners, whose opinions were sought on whether or not they considered these to be acceptable approaches to urban remodelling. Tables 7.1–7.4 indicate a broad measure of acceptance of facadism in certain circumstances.

To try and find some general themes in terms of local authority planners' opinions about facadism, a series of statements were presented as part of a questionnaire. Those considering the statements were given a choice of agreeing or disagreeing with each statement. The statements and responses were as follows:

1 Facadism is sometimes an acceptable approach to redevelopment because:

(a) it retains the character of townscapes while at the same time allowing the provision of modern accommodation.

Agree	*Disagree*
71%	28.5%

(b) it enables the economic and social benefits of urban remodelling and conservation to be allied.

Agree	*Disagree*
14%	57%

(c) landmarks and symbols in townscapes are retained.

Agree	*Disagree*
71%	28.5%

(d) historical associations are retained.

Agree	*Disagree*
57%	28.5%

2 Facadism is generally an unacceptable approach to redevelopment because:

(a) interiors of buildings should not be divorced from exteriors.

Agree	*Disagree*
71%	28.5%

(b) valuable historic fabric behind the facade is lost.

Agree	*Disagree*
85%	14%

(c) urban conservation is trivialised.

Agree	*Disagree*
28.5%	71%

(d) false environments are created and townscapes become merely stage sets.

Agree	*Disagree*
85%	0%

(e) it interferes with the proper evolution of modern buildings and a historical townscape continuum.

Agree	*Disagree*
57%	42%

(f) undesirable social and economic changes can result.

Agree	*Disagree*
42%	42%

(g) it changes townscape grain to the detriment of urban character.

Agree	*Disagree*
42%	42%

3 Facadism is acceptable as a last resort to retain street scenes where the rest of buildings cannot be saved.

Agree	*Disagree*
100%	0%

4 Facadism should not occur. Either a building should be refurbished, as far as possible in its existing form, or demolished and replaced with a new building in a contemporary style.

Agree	*Disagree*
71%	28.5%

5 Facadism is inappropriate for buildings of certain eras or types of construction, e.g. timber frame.

Agree	*Disagree*
28%	70%

6 Facadism is an element of postmodernism.

Agree	*Disagree*
42%	42%

The attitudes towards concepts such as facadism held by different interest groups is a huge subject which cannot be adequately covered in a book such as this. The purpose here has not been to prove any particular points or truths, but to give an insight into some of the relevant prevailing views within an influential part of the UK planning system and to provide background material to the case studies that follow. These views will, however, be of general interest to some readers.

Implicit in a perspective that regards the attitudes of public-sector planners as fundamental to development outcomes, is a top-down or policy-centred view of the policy formulation and implementation process, wherein policy decisions are seen as the starting point and are equated with action that is assumed to follow naturally from declared policy intentions. There is also the implicit notion of hierarchical organisation with policy emanating from the top (for example, central government or, at local level, planning authority committees), and being transmitted down through the decision-making process to the implementing agencies at the bottom of the hierarchy; in the case of building projects this will be the developers who, in this scenario, are seen to be in compliant relationships with the planning authorities. However, in many cases, implementing agencies are autonomous private-sector organisations with their own policy-making role and objectives, which may be in conflict with those of local authorities. If so, the developers will attempt to resist, overcome or amend the policies of other interests that are in conflict with their own.

The bottom-up perspective adopted in the case studies has focused attention on the various power relations and the inherent aims and values of different agencies within decision-making processes that have led to particular outcomes. This approach recognises that action or inaction may not relate to specific or explicitly stated policies. A particular decision may be a response to pressures or problems experienced 'on the ground'. For example, an expressed intention by planners to promote the retention of a particular facade may be the result of a building being in danger of collapse and the facade, which is important in townscape terms, is the only part of the building which is restorable. Situations such as this, encountered during the course of development projects, may have the effect of changing or refining previously formulated policies. Further, the analysis recognises that substantive issues such as facadism are not merely technical issues that exist in isolation from the contemporary development process, but are subject to political, social and economic pressures. The case studies therefore focus on managerial aspects of the relationship between facadism and the processes of conservation and heritage planning.

Case Study 1 – Nos 1–6 Brunswick Square, Bristol

The development of this site was completed in 1984, which was 23 years after discussions first took place between the local authority and a developer concerning the comprehensive redevelopment of the area and implementation of the Council's road proposals of the time. The scheme involved the construction of a four-storey contemporary office building fronting Bond Street, York Street and Gloucester Street, which incorporated the retention of most of the original facade of Nos 1–6 Brunswick Square. Parts of the facade had to be rebuilt because the original fabric was so badly decayed. The former side elevations and part of the roof form were also rebuilt. All the rest of the original fabric had either collapsed or had to be removed because of dry rot and decay, and attempts to persuade the developers to incorporate stub walls, to partly reflect the former subdivisions behind the facade, failed. The new building has a staggered elevation to Bond Street (see Plate Section, xx) and is finished in golden buff brickwork, a somewhat alien material to Bristol townscape, with bronze-coloured glazing and window frames. Sadly, the building is a bland and uninspiring composition and a disappointing contemporary companion to the elegant Georgian development nearby.

Historical, architectural and townscape context

Conceived in 1766 as a grand uniform square and part of a plan to build an elegant and fashionable residential district, Brunswick Square was never completed. However, along with the nearby Portland Square, which was completed in the nineteenth century, it still represents a rare and important example of Georgian urban planning on flat ground in Bristol. The heyday of this part of the city was near the end of the nineteenth century, when it was a fashionable residential district. By the 1930s it had declined in popularity and eventually became an inner city industrial and warehousing sector. Some buildings became disfigured externally by advertisements and inappropriate structural alterations, and interior dividing walls were sometimes removed to create larger working spaces. War damage also took its toll in the area. Subsequently, further economic and social change led to buildings being left empty and a survey carried out by the City Planning Department in the late 1970s indicated that the Portland Square area contained the greatest amount of dereliction in the city centre.

4 Facadism should not occur. Either a building should be refurbished, as far as possible in its existing form, or demolished and replaced with a new building in a contemporary style.

Agree	*Disagree*
71%	28.5%

5 Facadism is inappropriate for buildings of certain eras or types of construction, e.g. timber frame.

Agree	*Disagree*
28%	70%

6 Facadism is an element of postmodernism.

Agree	*Disagree*
42%	42%

The attitudes towards concepts such as facadism held by different interest groups is a huge subject which cannot be adequately covered in a book such as this. The purpose here has not been to prove any particular points or truths, but to give an insight into some of the relevant prevailing views within an influential part of the UK planning system and to provide background material to the case studies that follow. These views will, however, be of general interest to some readers.

Implicit in a perspective that regards the attitudes of public-sector planners as fundamental to development outcomes, is a top-down or policy-centred view of the policy formulation and implementation process, wherein policy decisions are seen as the starting point and are equated with action that is assumed to follow naturally from declared policy intentions. There is also the implicit notion of hierarchical organisation with policy emanating from the top (for example, central government or, at local level, planning authority committees), and being transmitted down through the decision-making process to the implementing agencies at the bottom of the hierarchy; in the case of building projects this will be the developers who, in this scenario, are seen to be in compliant relationships with the planning authorities. However, in many cases, implementing agencies are autonomous private-sector organisations with their own policy-making role and objectives, which may be in conflict with those of local authorities. If so, the developers will attempt to resist, overcome or amend the policies of other interests that are in conflict with their own.

The bottom-up perspective adopted in the case studies has focused attention on the various power relations and the inherent aims and values of different agencies within decision-making processes that have led to particular outcomes. This approach recognises that action or inaction may not relate to specific or explicitly stated policies. A particular decision may be a response to pressures or problems experienced 'on the ground'. For example, an expressed intention by planners to promote the retention of a particular facade may be the result of a building being in danger of collapse and the

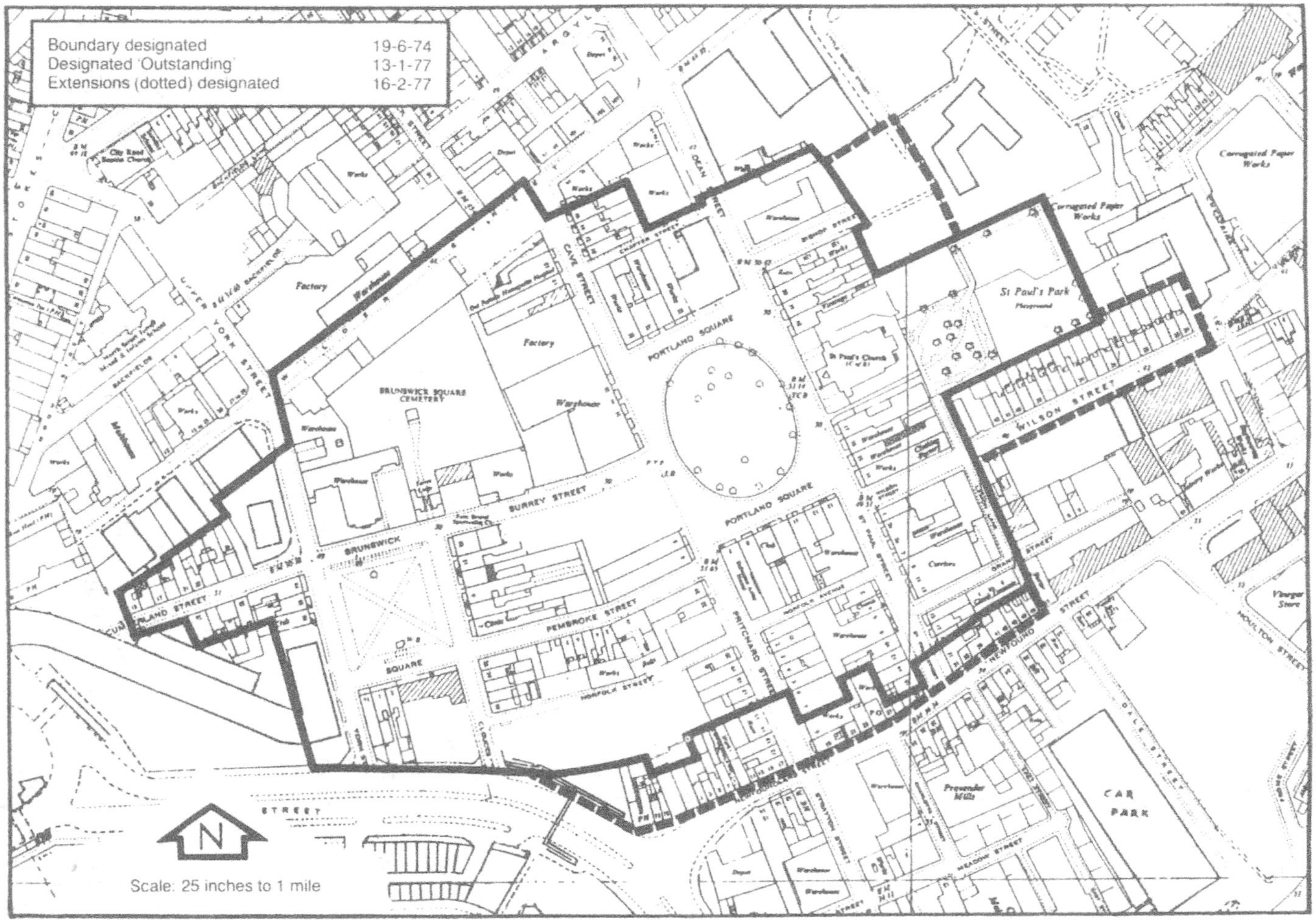

Plate 14 Portland Square conservation area, Bristol.

The role of the Secretary of State

The buildings were saved by the advent of legislation requiring listed building consent to be obtained for the demolition of listed buildings. In 1972, following considerable lobbying by both local and national amenity groups, the Secretary of State for the Environment called a public inquiry into the proposals affecting Nos 1–6 Brunswick Square. It appears that the following points were of particular significance to the Secretary of State's consideration of the proposed development:

1 The importance of the buildings both intrinsically and in relation to their context.
2 Their structural condition and state of repair.
3 The likely cost of repairing and maintaining the buildings in relation to their importance, with due regard to their economic value when repaired.
4 The possibility of incorporating the buildings in any proposed redevelopment of the overall site.

Following the public inquiry the Secretary of State subsequently intervened to overrule his appointed appeal Inspector and refused consent for the demolition of the buildings. The document giving the Secretary of State's decision indicated that:

> in spite of its present derelict condition, the building could be incorporated into the redevelopment proposals and that in view of its importance to the square and the fact that only two of these Georgian Squares on flat ground remain in Bristol, it should be preserved. Permission to demolish the premises could not be justified purely in order to secure the uniformity of height with the new office buildings along the northern side of Bond Street. In his view, the appearance of Brunswick Square should be both maintained and enhanced and he cannot agree that there is no option but to allow the scale and character of the square to be disrupted by permitting listed building consent to demolish Nos 1–6 in order to allow it to be substituted by a new office block.
>
> (Decision letter of the Secretary of State for the Environment to Bristol Corporation, 14 November 1972)

Missed opportunity

In the light of the outcome in this case, when only the facade and side elevations of Nos 1–6 Brunswick Square were restored or replaced, and the eventual efforts to save the buildings in their entirety failed (see Plate 15), it is ironic to find that a developer had approached the local authority in 1972 and again during the early part of 1973, with a view to refurbishing the historic buildings as part of a development of small office suites. At the time the Council were awaiting the outcome of a challenge in the High Court of the Secretary of State's refusal of listed building consent to demolish Nos 1–6.

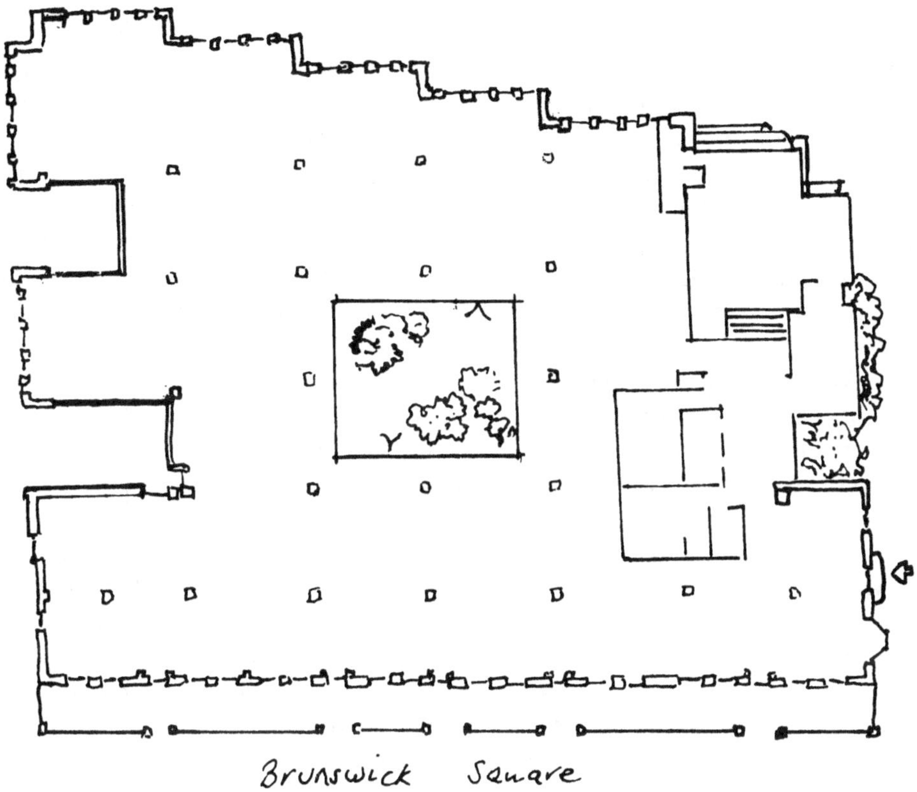

Plate 15 Author's sketch, ground-floor plan of the implemented scheme involving Nos 1–6 Brunswick Square, Bristol, after original by Towning Hill & Partners, architects.

Consequently, no encouragement was given to the developer, who eventually lost interest in the scheme. The Council's preoccupation with their plans for comprehensive redevelopment in the area, even in the face of mounting public pressure for the retention of the city's historic townscape, can be seen as an important causal factor in facadism being an outcome in this case. Had the commitment to restoring the city's historic built environment that emerged within the local authority in the mid-1970s been present a few years earlier, it is likely that Nos 1–6 Brunswick Square would have been restored in a more authentic form than that which was eventually forced on the City Council and the local community.

Commitment to conservation

The Secretary of State's decision to prevent the demolition of the listed buildings can be seen as a watershed in this case. The rebuff for the authority's and prospective developer's aims of continuing the comprehensive redevelopment in the area eventually set a new tone for negotiations over the next twelve years, wherein the preservation of the southern side of the historic Brunswick Square became a fundamental concern.

During the 1960s, local amenity groups in Bristol had begun campaigning against the attitude of the planning authority in relation to numerous proposals. They gained great kudos in first orchestrating objections to proposals affecting the Grand Spa Hotel, which forced a public inquiry, and then gaining national press coverage following their influential arguments presented to the inquiry which resulted in the minister at the time, Peter Walker, rejecting the proposals.

It is noticeable from local authority records that conservation interest groups were expressing concern in the very early 1970s that the buildings at Nos 1–6 Brunswick Square were severely deteriorating. The first signs of a commitment on the part of the local authority towards conserving the historic townscape of Brunswick Square emerged in 1974, when it was included within the newly designated Portland Square Conservation Area. In 1975, the same developer who had worked with the local authority to bring about the St James Development returned to discuss a scheme involving the retention of Nos 1–6 Brunswick Square. Proposals were drawn up in close consultation with the City Planning Department, which indicated the listed buildings were to be repaired and refurbished and provided with a new mansard roof in secondhand pantiles to match other buildings in the square. The rear wall of the building was to be demolished and rebuilt in facing brickwork and the side elevations altered, but in keeping with the facade. The new block fronting Bond Street was to be clad in red sandstone colour clay tiles. In February 1976, the scheme was amended to show the rear elevation of the listed buildings removed and various other changes to the design of the new building. This scheme floundered, however, following the collapse of the property market.

The appointment of a new City Planning Officer in 1976 gave further momentum to the emergence of the commitment to conservation in the city. In terms of the Portland Square area, this was further emphasised in 1977 when outstanding status was conferred and funds set aside to enhance the area under the newly formed Conservation Programme. The Historic Buildings Council (HBC) had declared Bristol a 'Priority' city, a status reserved for places of the highest historical importance with large numbers of listed buildings. This gave the city high priority when grant aid was allocated and for 1977–78 the HBC matched the city's input of £100,000.

A study made in 1977 by the planners confirmed that Bristol had a vast problem of dereliction among its historic fabric in the inner city. Given the scale of the problem the availability of grant aid was an important factor in achieving the retention of valuable townscape. Not only was aid available to help stimulate appropriate redevelopment or refurbishment schemes, but also to assist towards funding urgent remedial works for buildings in danger of collapse or studies into how best to preserve historic fabric. In the case of Nos 1–6 Brunswick Square, funds were used to carry out a restoration study which resulted in an unimplemented planning permission being granted in 1978 to the developer who had been involved with the site since the 1960s. This scheme provided for the refurbishment of the listed buildings which, as in the 1975 schemes, were to be linked to a new building fronting Bond Street. Also a 50 per cent grant was provided in 1979 towards essential works to stabilise the facade and remove masonry which was a danger to the public. In 1981 a substantial grant-aid offer to the eventual developer was also regarded by conservation officers involved in the protracted negotiations surrounding this case as a most significant factor in the retention of a vitally important but severely decayed townscape element. However, it is also highly significant that, during the four years prior to the redevelopment scheme, the strenuous efforts of local authority officers to obtain general grant-aid funding did not encourage the developer, who at the time had been involved in various schemes for the site over a period of twenty years, to implement the existing planning permissions. Clearly, at that time market indicators were not conducive to an economically successful scheme.

Market influence

In February 1975, when market conditions for modern office space in this part of the city were still relatively acceptable to developers, planning and listed building consent applications were submitted to the local authority relating to the refurbishment of the facade and side elevations of Nos 1–6 linked to a new office building behind. Difficulties in reaching agreement with Avon County Council concerning the amount of land required to provide a suitable footpath along Bond Street led to delays and the submission of an amended scheme. By the time that a favourable decision had been reached by the City Council to advise the Department of the Environment that it was disposed to grant consent, the property market had crashed, leaving over one million square feet of completed office space in Bristol, much of which was vacant. Consequently, the proposed development was abandoned. In 1978 the same developer returned with an application for the approval of reserved matters relating to the 1975 permission, which was subsequently approved but not implemented. The plans indicated more authentic proposals for the refurbishment of the historic building with the old and new elements being in essence separate buildings and greater subdivision inside the shell of Nos 1–6 Brunswick Square (see Plate 16).

It was after this consent that grant aid was offered by the HBC to assist the stabilisation of the old facade. There was apparent concern among Council officers about the developers' intentions with regard to the stabilisation works, and that having taken down some of the facade to make it more stable the developers might pull out of the proposed scheme, leaving a half-demolished building. Staff in the City Planning Department were attempting to co-ordinate the timescales of a building agreement with the developer, building regulation approval and the essential deadlines for grant-aid

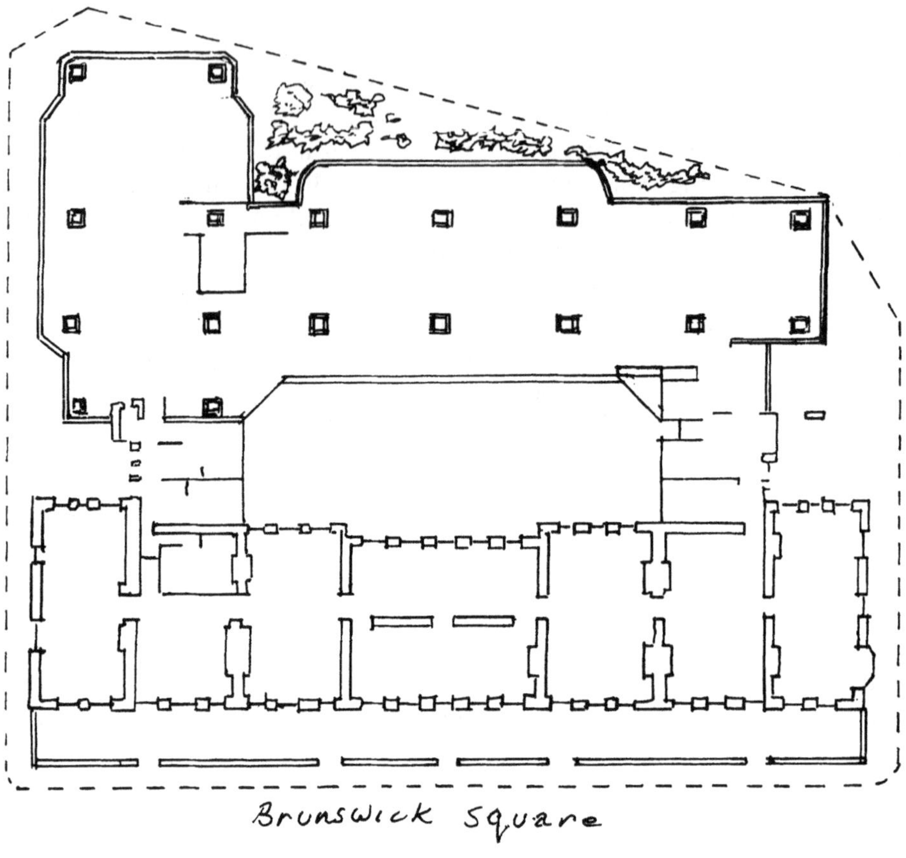

Plate 16 Author's sketch, ground-floor plan of the 1978 scheme relating to Nos 1–6 Brunswick Square, Bristol, after original by Towning Hill & Partners, architects.

awards from the HBC and the local authority. However, the developer clearly did not wish to proceed with the scheme because matters remained in abeyance until early 1980, when the developer, clearly encouraged by a revival of the market in office accommodation in Bristol, approached the planners to discuss further modifications to the proposed development. These involved the removal of the rear elevation of the listed buildings and a totally open-plan office layout. This approach was rejected at that time by the planners.

Dereliction

Dereliction was a critical factor that led to facadism in this case. Typical of the buildings of the period, much of the structural support at Nos 1–6 Brunswick Square was provided by timber members. When such members fail, the result is often the collapse of masonry at the same time or soon after. The removal of lead gutters at Nos 1–6 had caused long-term damp penetration, resulting in extensive damage through dry rot to structural timbers and masonry. Vandalism had also taken its toll and, by 1979, the building was dangerous to enter. By the following year the total collapse of the remaining fabric was imminent. The developer who eventually developed the site was, as a result of the critical condition of the historic fabric, able to resist a complete refurbishment scheme because the planners were desperate to see a scheme implemented which, as a last resort, would save the facade to Brunswick Square.

Impact of highway proposals

Another significant issue in this case was the impact of highway proposals. In the 1960s it had been envisaged that much of the land to the rear of Nos 1–6 Brunswick Square would be required for a road junction. However, after local government reorganisation in the 1970s, Avon County Council became the Highway Authority and devised a road scheme requiring less land, allowing greater scope for a new development behind Nos 1–6 Brunswick Square. Therefore, the notion of combining the historic fabric with a new building was born and the basis of the eventual outcome created. Despite the apparent lessening of highway constraints, a new development scheme involving Nos 1–6 Brunswick Square was not immediately forthcoming. While the Highway Authority now required less land there was still some uncertainty over the exact route of the road improvement line, which consequently undermined negotiations between Bristol City Council's planning and conservation officers and prospective developers. Clearly, delays in the negotiation process were to result in further serious deterioration of the building's structural condition. Eventually, however, by 1975 Bristol's planners were able to persuade Avon County Council to commit itself on the road line, which allowed negotiations on a scheme to save as much as possible of Nos 1–6 Brunswick Square to proceed.

Design and townscape issues surrounding the 1981 scheme

The developers were committed to building large-scale open-plan office accommodation to reflect projected market demand. It is clear from discussions with conservation officers involved with this case at the time, that their negotiating position in terms of achieving a more authentic interior layout was undermined because of the very poor structural condition of the remaining historic fabric. They were, however, able to negotiate the inclusion of features aimed at recreating the feeling of a traditional Georgian exterior with chimneys and party walls at roof level supported by steel beams within the roof space, and the provision of dormer windows (see Plate Section, iv). These features and the existence of the party quoins means that superficially the facade reads as separate town houses from the street. The lack of internal subdivision is not so noticeable by day, but during evening time the fluorescent strip lights destroy the illusion.

As far as the Council's conservation officers and the local amenity societies were concerned, the relationship of the proposed new building to the preserved facade and the impact of any proposed development on views from within the square were of considerable importance and a great deal of negotiating time was spent on these aspects. The new building does not interfere with views over the roof of Nos 1–6 Brunswick Square, but the way the new building links with the retained fabric was criticised by amenity societies and the City Council's Conservation Advisory Panel, a non-elected group which is supposed to provide informed lay opinions on proposed development. The Panel, while recognising the urgent need to proceed with the refurbishment of the facade to Brunswick Square, felt that the scale relationship of the new to old fabric was very poor and that the rebuilt side elevations to the old buildings should be finished in dark rather than light coloured render (see Plate Section, v). The planning officers' recommendation, which was accepted by elected committee members, was that the scheme under consideration at the time was acceptable given the need to act quickly to save as much as possible of Nos 1–6 Brunswick Square.

Policy perspective

The notion of facadism was explicitly accepted in the original conservation policy document for the area published in 1974 which stated: 'Where sites form an essential part of the street pattern, redevelopment should involve the rebuilding of facades so that the original street pattern may be restored.'

In relation to Portland Square, which was considered the finer example of the two squares, the document recognised that the 'facades and general scale must be rebuilt or retained if the original unity of the square is to be kept'. An enhancement scheme document, which followed in 1979 after the conservation area was given 'outstanding' status, contained no explicit references to facadism but expressed objectives such as the restoration of the squares to their original form and the construction of replica buildings and sympathetic frontage development in terms of form, massing and materials. There is an apparent subtle change of emphasis in the planners' policy statements between 1974 and 1979 towards a more authentic reconstruction of the original townscape form. If the schemes negotiated between the developers and planners in 1975 and 1978 are analysed closely it is possible to detect this shift in emphasis being confirmed. In 1975, the planning officers had worked closely with the developer to produce a scheme which showed much of Nos 1–6 Brunswick Square restored and rebuilt including its rear and side elevations, with this element being linked to a new office building fronting Bond Street. However, the developer subsequently amended the proposals to omit the rear elevation of Nos 1–6 and the rebuilt fabric linked directly into a new open-plan office building of similar height to the retained fabric. To create the impression of the original form of the buildings, new party walls were shown extending back from the restored facade into the new office space for a distance of under half the depth of the original buildings.

In 1978, a greater degree of restoration is displayed. Again the planners worked closely with the developer's architect to produce an appropriate conservation scheme and initiated a grant-aided study of how the buildings could best be conserved. The result was a scheme which indicated internal walls, fireplaces and the rear elevation restored and, arguably, more sensitive link blocks between old and new fabric. While the 1978 scheme was in essence similar in many respects to the first scheme produced in 1975, it was a more authentic representation of the form of the original buildings, although it still involved linking the historic fabric to a contemporary office building. Also, the second scheme agreed by the planners in 1975 involved the omission of the rear elevation of Nos 1–6 Brunswick Square, which represented a reduction in the commitment to retain an authentic version of the historic buildings. It also introduced the notion that facadism would be an acceptable conservation approach in this case. An important difference in the negotiations between planners and developers in 1975 and 1978 was that, after the instigation of the Bristol Conservation Programme in 1977, the Brunswick Square scheme had been allocated grant-aid funding, which gave the planners greater scope to encourage developers to produce development proposals that complied with their objectives. The advent of financial provisions to back up statements of intent represented a greater policy commitment to authentic restoration.

However, the eventual outcome did not achieve the level of authenticity envisaged by the planners in 1978, despite the encouragement of grant aid. The decision to restore just the facade and side elevations of the listed buildings, with no interior features included, was a pragmatic response to situations 'on the ground'. The alternative was the collapse and possible complete loss of the remaining traditional townscape. The planners' policies in relation to this townscape ranged from explicit statements in 1974 that the retention of facades was fundamentally important to the character of the townscape, to the inclusion of Nos 1–6 Brunswick Square in the Conservation Programme in 1978 when, through the use of public funding, the private sector was expected to restore most of the form and many of the interior features of the original buildings to, finally, a desperate attempt to save simply the facade and side elevations of the listed buildings. Rather than facadism being an outcome dictated by planning policy decisions in this case, it appears that the eventual policy to adopt facadism was tailored by factors such as the condition of the buildings, the state of the property market and the aspirations of the developers.

Summary

There are a number of points which this case highlights in terms of facadism as an outcome in the urban development and conservation processes. The attitude of the local authority towards its historic townscape during the 1950s and 1960s was fundamentally significant. The potential value of the buildings was completely disregarded at this time and, consequently, they fell into serious disrepair. A lack of monitoring of their condition over many years and a failure to instigate measures to prevent further deterioration, until the collapse of what remained of the buildings was imminent, placed the planners in an impossibly weak negotiating position by 1981, at which stage the compromise solution of facadism became almost an inevitable outcome.

Also, the case demonstrates the importance of development proposals being in harmony with market conditions in capitalist economies. The timing of the proposals in 1975 and 1978 was unfortunate, as the property market was in recession. Had conditions been more economically favourable, a more authentic conservation scheme might have been achieved.

The outcome can be seen as the result of a complex interplay of factors and conflicting interests and one wherein various objectives had to be compromised. The developers, for instance, obtained a marketable open-plan office development in a favourable location. Conservation interests achieved the preservation of an important townscape element in the facade of Nos 1–6 Brunswick Square without which the formality of the Georgian composition would have been destroyed. The developers were constrained in terms of the floorspace they were able to achieve by conforming to the height of the retained facade and respecting the views out of the square to the south. It could be argued that the true character of Brunswick Square has been damaged as a result of facadism, in that on close inspection there is clearly a contemporary building lurking immediately behind the preserved facade and that the scheme lacks the necessary authenticity to reflect the true spirit of conservation.

Clearly, the role of certain actors and agencies in the decision-making processes influenced the outcome both directly and indirectly. The Secretary of State, for instance, was a key actor in this case in that his overruling of the appointed Inspector and the prevention of the demolition of Nos 1–6 Brunswick Square saved the facade which was eventually preserved. Had this intervention not occurred, the extension of the St James office development would almost certainly have covered the southern side of Brunswick Square. The amenity societies were also instrumental in preventing the 1960s proposals destroying the southern side of the Georgian square. The Highway Authority, whose main objective was not the protection of historic townscape but to ensure that sufficient land to the rear of Nos 1–6 Brunswick Square was available for their proposed road junction scheme, also influenced the outcome. First, the decision to reduce the size of the proposed road junction at Bond Street created the space for a development involving the retention of part of the listed buildings and the construction of a new building facing Bond Street. Without this proposal change by the Highway Authority, a totally different outcome would have evolved. Second, the delays caused by the uncertainty over the highway proposals in 1975 possibly contributed to the abandonment of what would have been a more authentic conservation scheme than that eventually built.

Finally, the case demonstrates that despite the framework of policy statements concerning the historic built environment, a particular outcome such as facadism may be more of a reflection of situations 'on the ground' than a policy directive filtering down from the top of a power hierarchy. Also pragmatic solutions such as facadism can be seen as refining influences in the process of conservation policy formulation. Policy stances change as a result of particular situations and circumstances. For example, with Nos 1–6 Brunswick Square, an aspiration of conservation planners to restore the listed buildings was modified to an attempt to save simply the facade. Also the outcomes in such cases provide examples against which future decisions and strategies can be examined.

Case Study 2 – Brunel House, St George's Road, Bristol

In 1981, Co-operative Insurance Society (CIS) commenced the redevelopment of what was originally an impressive hotel, dating from 1837, designed in collaboration with Isambard Kingdom Brunel to accommodate passengers travelling to America on his Great Western Railway and Steamship Companies. The building had fallen into decline after the Second World War and was the subject of numerous development proposals involving its demolition. The CIS scheme preserved the building's magnificent facade and an extraordinary feature called the Horse Bazaar, a spacious yard enclosed by cliff-like brick walls built into the adjacent hillside (see Plate 17). This feature, built by a William Rogers in about 1839, provides a fascinating link with the days of horse-drawn carriages and trams. It was an auction place where the gentry came to see bloodstock paraded. The yard has been transformed into a sort of amphitheatre, which retains the towering brick wall stretching up to the gardens of Great George Street at the back of the former auction ring. The theatrical setting is enhanced by the traces in the wall of the original ornamental archways. These, it is said, once led to the stables which were reached by ramps.

Brunel House today

The new building behind the facade has three sections with an open landscaped court in the centre. Inside, the accommodation is mainly open-plan, which provides reasonably good quality office space, although the area at the front of the building under the colonnade tends to be poorly served by natural light. Imaginative use has been made of raised levels at the back of the site which provide interesting and attractive landscaped terraces. The western side elevation of the original building collapsed after the fabric behind was removed and has been rebuilt in Bath Stone to a slightly different style. The new building was constructed in reddish-brown brick which blends relatively successfully with the Bath Stone facade, the walls at the back of the site, the Horse Bazaar and other buildings in the locality. From St George's Road, in close proximity to the building (see Plate 18), the roof form is not visible. Descending Brandon Steep on the

Plate 17 The Horse Bazaar, retained and landscaped within the development of Brunel House, as seen from St George's Road.

western side of the building and from other vantage points, however, the roof form appears monolithic and over-dominant. This is the most disappointing aspect of the scheme, and illustrates the importance of carefully considering roofscape when facadism is to be a chosen development option (see Plate 19 and Plate Section, vi and xix).

Facadism becomes official policy

In 1976, adopted policies were included in a guidance scheme for St George's Road, which consolidated the pressure from outside interests and gave formal recognition to the Council's aspiration to retain the facade of Brunel House. CIS seemingly accepted the need to incorporate the retention of the historic facade into their proposals for the

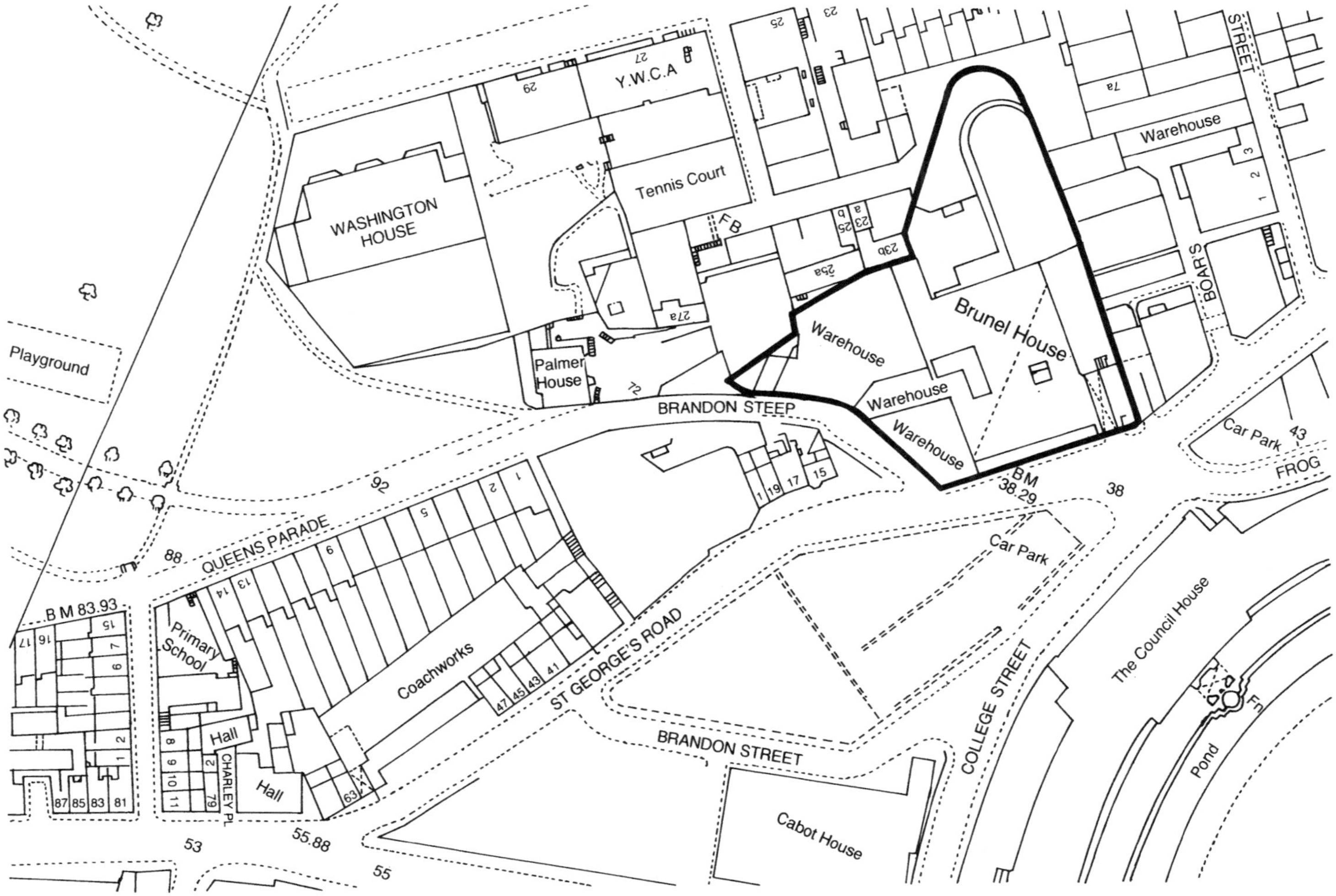

Plate 18 St George's Road, central Bristol.

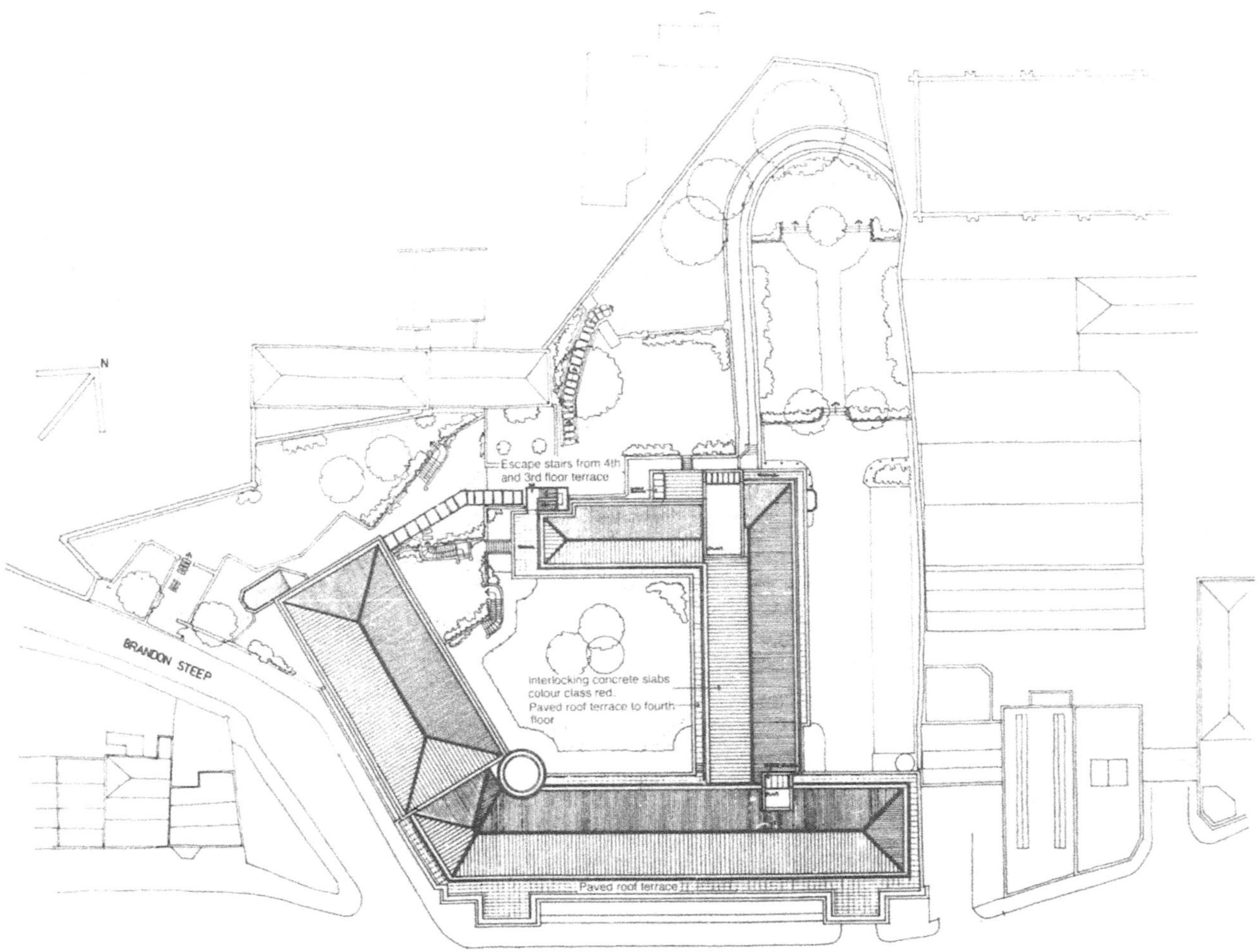
N
Escape stairs from 4th
and 3rd floor terrace
BRANDON STEEP
Interlocking concrete slabs
colour class red.
Paved roof terrace to fourth
floor
Paved roof terrace
ST. GEORGES ROAD

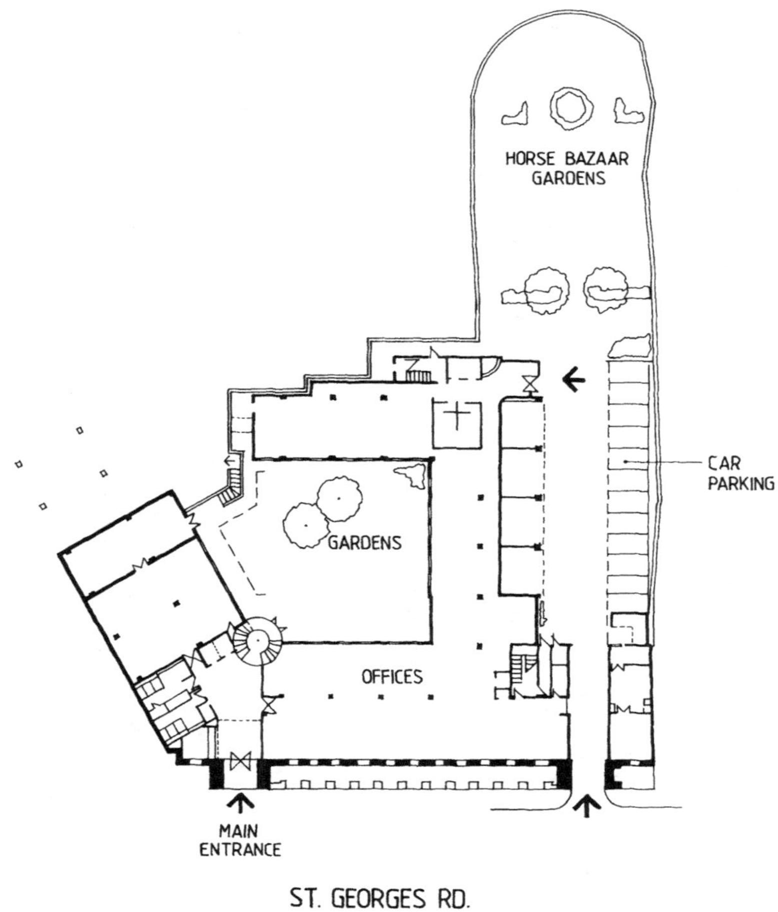

Plate 19 Ground-floor and roof plans, Brunel House, Bristol, with the kind permission of Alec French Partnership, architects, and Bristol City Council.

Brunel House site, because in 1976 a scheme was submitted to the city council which fully reflected the St George's Road guidance scheme.

The market and planning gain

Two factors led to the abandonment of the 1976 proposals submitted by CIS. First, the state of the property market; second, the fact that the development of Brunel House was linked by a planning agreement to another site which was to be used to provide inner-city housing. Originally, the city council housing department had expressed an interest in acquiring the completed housing, but by 1977 it became clear that the authority was in no position to do so. Eventually, a housing association became involved in protracted negotiations between the developers, the architects and the planners, which extended over the next four years.

The planners were using Brunel House as a lever to extract further conservation gains in the area. Developers were keen to proceed at Brunel House, because by 1979 the previously depressed office market was regenerating. The local authority had imposed restrictions on new office space unless conservation gains were involved, and the retention of the facade of Brunel House offered a means of achieving a relatively rare new office building behind the historic facade. CIS did not feel able to commit themselves to a series of schemes along St George's Road which the planners hoped would clear up the extensive dereliction in the locality. An impasse was reached, with CIS claiming that they did not have the funds to develop any other sites in the area apart from the Brunel House site and the planners refusing to countenance the development of Brunel House in isolation.

The complex negotiations surrounding the Brunel House site eventually produced a scheme in 1979 which involved CIS entering into a legal agreement to sell their site (which had been earmarked previously for housing development) to a Housing Association prior to the development of Brunel House taking place. However, the legal agreement remained unsigned, because CIS experienced land assembly problems regarding a pocket of land that was required to enable development of the housing site.

Questions of viability

Another dimension to the negotiations was the developer's concern over the escalating cost of restoring the historic facade of Brunel House and the impact of this on the viability of the project. Consequently, the developers approached the City Council to discuss the possibility of receiving grant aid under the city's conservation programme. The failure to bring about the implementation of a development scheme which would ensure the future of the facade of Brunel House was worrying the planning authority and investigations were made into the availability of funds for grant aid. However, given other commitments within the conservation programme, the authority felt unable to offer any funding. In a desperate attempt to facilitate development, the planners resolved a strategy to stimulate action. They suggested to the developers that unless some development occurred along St George's Road in the near future the whole situation would have to be reviewed, with the prospect of much more onerous policies being formulated by the planning authority. CIS responded with the submission of a development scheme for Brunel House alone, comprising the preservation of the classical style facade, construction of 79,000 square feet of office accommodation behind, and the restoration and landscaping of the interesting open yard set into the hillside and surrounded by high retaining walls, which originally comprised the Horse Bazaar. The proposals were approved subject to the completion of a planning agreement which would ensure that the CIS land holding earmarked for housing development would be sold to a housing association before Brunel House could be redeveloped. Following its approval the planners adopted a development brief for St George's Road, which stated that any subsequent applications or renewals for development of the Brunel House site would only be considered if accompanied by proposals for the development or disposal of the developers' other land holdings in the locality. The combination of this strategy and the buoyancy of the office market in Bristol during 1981 convinced CIS to proceed with the redevelopment of Brunel House without the assistance of grant aid. Interestingly, the completed building was subsequently occupied by the city planning department and the officers who had worked so vigorously to save Pope's magnificent facade.

Summary

As in the previous case, a historic facade was only saved from demolition by chance. CIS decided to try and assemble a large area of land along St George's Road, to bring about comprehensive redevelopment rather than adopt an incremental approach. Had the latter occurred the valuable townscape feature of the facade to Brunel House would probably have been destroyed and a modern tower block or other anonymous building constructed on the site.

Facadism became the most likely development option in the case of Brunel House because of the unease among some planners about the rising tide of public and amenity group opposition to the possible demolition of the historic building. Also, local government reorganisation in the early 1970s had a considerable impact in this respect, with several of the elected members (who had been extremely pro-active in terms of promoting the construction of modern commercial buildings in Bristol) moving to serve on Avon County Council. Even during the days of Modernism and rampant commercial exploitation of the city's built environment, facadism had been on the agenda, with informal advice being given to prospective developers by planning officers to the effect that the preservation of the facade of Brunel House would be likely to overcome the considerable objections being received by the local authority from amenity groups and the public. Facadism may be regarded in some circumstances, as shown here, as an instrument of compromise whereby developer aspirations and public opposition can be reconciled.

The case also reveals that facadism provides a means of overcoming private-sector reluctance to the updating of historic buildings which have become obsolete and are considered unmarketable. Even though Brunel House survived the post-war commercialist policies of the local authority, its poor structural condition and perceived obsolescence rendered it an unlikely target for private-sector-funded refurbishment.

Facadism is seen here as a pragmatic strategy by the local authority to achieve the preservation of a highly important townscape feature. Eventually the developers also regarded facadism as a pragmatic approach to unlocking the development potential of the Brunel House site and providing them with a marketable modern office development. Of all their land holdings along St George's Road, Brunel House presented the most attractive development option. With this in mind, the planners' strategy was to use the Brunel House site as a lever to achieve other conservation gains in the locality. Had they insisted on full refurbishment of Brunel House rather than facadism, it is unlikely that such a strategy could have been employed because the Brunel House site would then have been far less appealing to private-sector developers. The strategy resulted eventually in the release of the housing site referred to above, which has since been developed by a housing association in accordance with the planners' urban design requirements.

Finally, this case demonstrates that facadism may sometimes be a means of saving historic townscape elements through private-sector resources without recourse to public-sector funds. While CIS attempted to make use of the Bristol conservation programme in the restoration of the facade at Brunel House the local authority felt unable to divert resources from other priorities, probably feeling that the Brunel House scheme was sufficiently marketable to the private sector to be able to save the facade of Brunel House without the use of public funds. The Brunel House case therefore illustrates that facadism, which combines the retention of important townscape elements with the creation of highly marketable accommodation, can provide a valuable pragmatic solution to conservation problems, given the finite supply of public funding.

Case Study 3 – Nos 4 and 6 St Andrew Square, Edinburgh

During 1992 the Scottish Provident Institution (SPI) were in the process of extending their existing headquarters at No. 6 St Andrew Square, Edinburgh to encompass Nos 4 and 5. The development involved the retention of the facade and front rooms of No. 4 and just the facade of No. 5. The development is the outcome of a process which began in 1976 when SPI submitted proposals for expansion and were refused planning permission for the redevelopments of Nos 4 and 5 St Andrew Square, involving demolition and new build. The facade retention scheme has involved a considerable amount of construction and refurbishment. Defective sections of stone have been replaced in both facades, particularly at eaves cornice level at No. 5. Pitched roof profiles have been reconstructed similar to those on the former buildings and finished in natural slate. In the case of No. 4, the scheme involves the reproduction of original porches constructed in similar stone and the retention of internal cornices, panelling, fireplaces and timber features of the main front rooms. Also included are the reconstruction of an ornate chimney head, the infilling of the doorway with a glazed panel on the original door line, and the restoration and refitting of an iron gateway at No. 5.

The 1992 scheme was the outcome of a decision by elected committee members of Edinburgh District Council to disregard the advice of their professional planning officers. It had been recommended that Nos 4 and 5 St Andrew Square should be demolished and replaced by a new building of high-quality design, which would have enhanced the symmetry of the square and created a better townscape relationship between SPI's existing Modernist building erected in 1968 and the historic building at No. 3 St Andrew Square.

Historic, architectural and townscape context

St Andrew Square is an important townscape feature within Edinburgh's First New Town; it lies at the eastern end of George Street and balances Charlotte Square at the other end (see Plate 20). The character of St Andrew Square lacks formality because of the diversity of building styles and proportions. Nos 4 and 5 are part of the southern frontage to the square. SPI's Head Office at No. 6 St Andrew Square, and extending to Meuse Lane at the rear, is unashamedly modern in style. Designed by Rowand Anderson Kininmonth and Paul, the building is now regarded as an imaginative response to the problems of building a sunless north-facing elevation and, in British terms, it is held to be an exemplary building of its time. Nos 4 and 5 St Andrew Square are both listed buildings (Scottish listing category B). Of the two buildings, No. 4 is the most interesting architecturally. It is an Italianate building to the design of John McLachlan and dates from 1883. The lively symmetrical composition is rich in classical ornament and when first built would have reflected the domestic scale of St Andrew Square. Sadly it has since been dwarfed by neighbouring commercial buildings. No. 5 St Andrew Square is less distinguished, having a somewhat austere appearance and plain, utilitarian detailing.

In townscape terms the value of the facades of Nos 4 and 5 St Andrew Square rests in the architectural quality of No. 4 and that as examples of Victorian and Edwardian elements they help to maintain a variety of period architecture in the square. There is, however, a certain incongruity to the southern frontage of the square with No. 3 towering above No. 4, which is considerably lower in height than No. 5. Next to No. 5 is SPI's Modernist Head Office, which is making its own architectural statement without reference to history.

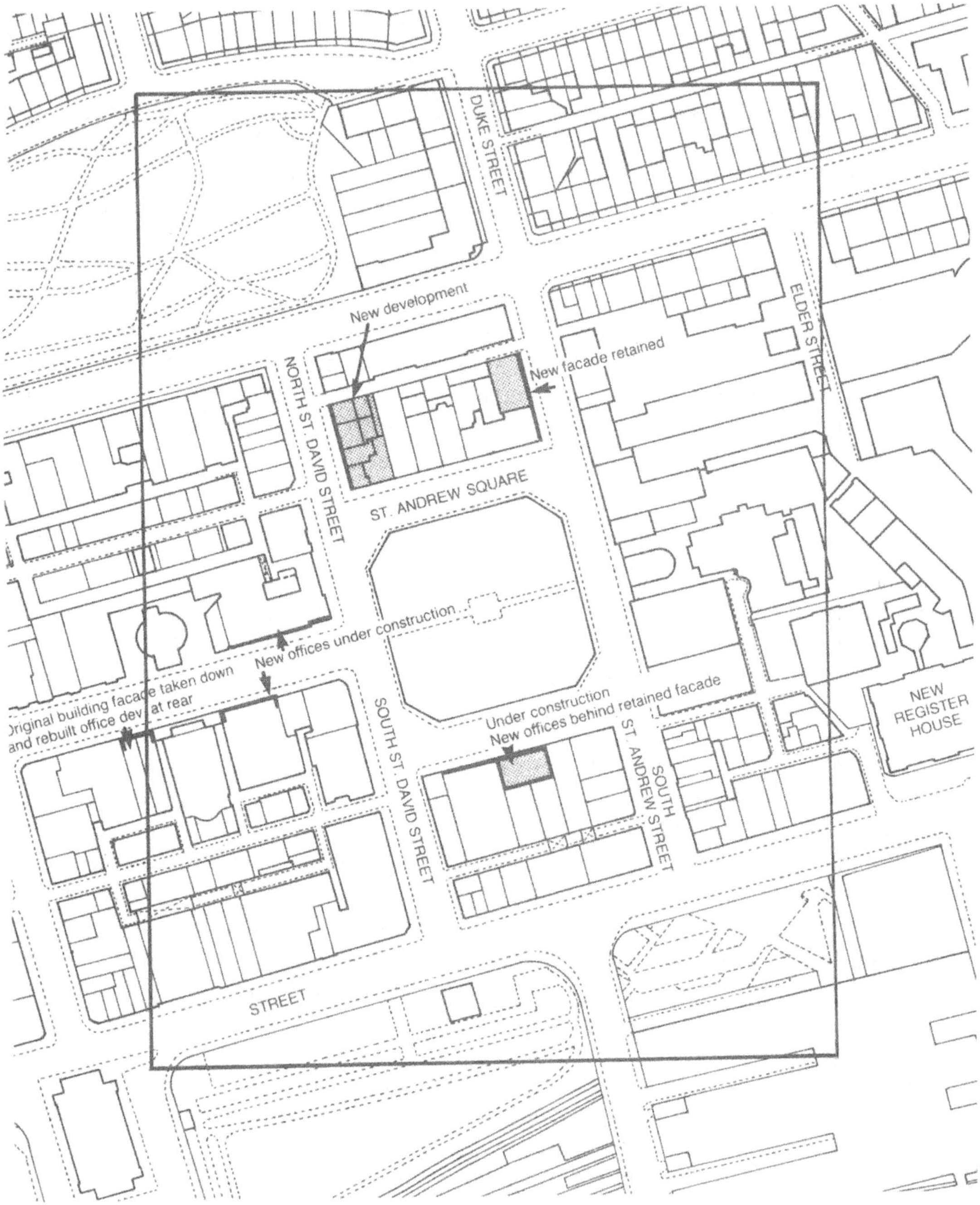

Plate 20 St Andrew Square, Edinburgh.

SPI's need to expand

The Scottish Provident Institution, a mutual life assurance office, has been based in St Andrew Square since 1841 and at No. 6 since 1863. Over the years the Institution has expanded towards Meuse Lane and west on to South St David Street. Nos 4 and 5 were purchased by SPI in 1974 in preparation for the expansion of their Head Office. Shortly after this, the buildings were statutorily listed. SPI had evolved an overall master plan to provide a new Head Office complex with the necessary flexibility to meet expanding and changing requirements. In particular, floor levels in each phase were fixed at the same level in the interest of efficiency. The proposed redevelopments in 1976 would have reflected this master plan. Having been refused planning permission for this redevelopment, one of the options open to SPI could have been the conversion of the existing listed buildings. However, the cellular nature of these buildings was considered completely inappropriate in the light of the Institution's need for flexible open-plan spaces. Having carried out an appraisal to determine whether or not the existing buildings could be linked together and to the existing Head Office at No. 6 in a manner that would provide sufficient space of the quality required and which would be adequately linked to the present Head Office accommodation to provide the necessary flexibility in use, it was concluded that:

1 The floor levels and the floor-to-floor heights of Nos 4, 5 and 6 are at such variance with one another that satisfactory office space could not be created.
2 In view of the disparity in floor relationships the requirements for continuous service zones, access floors and ceiling heights could not be achieved.
3 The necessary alterations that would have to be made in linking the buildings together in order to satisfy building regulations and also to provide the required ease of circulation would prove disproportionately expensive in relation to the floor areas achieved.
4 The financial constraints of converting the existing buildings place a limit on the total cost per square foot. Any solution involving the retention of the existing buildings would result in less usáble floor space, and costly increases in circulation space. Therefore, in comparison with a new build scheme, there are fewer funds available to devote to achieving the required quality of building.

In June 1987, SPI submitted a consultation document to the Director of Planning at Edinburgh District Council, which outlined the Institution's need for expansion. Increases in staffing levels, the amount of computers and other technology and the Institution's projected substantial further increase in business transactions over a ten-year period were in the late 1980s making it essential for SPI to expand their Head Office. The organisation was committed to remaining in Edinburgh rather than relocating elsewhere because, in its opinion, Edinburgh was fully capable of sustaining its distinctive role within the financial markets.

SPI's requirement for integrated development

In SPI's opinion, for their enlarged Head Office complex to meet the needs of the Institution during the 1990s and beyond, the next development phase should be fully integrated with the existing complex. Integrated development should provide consistent floor levels to allow easy communication, avoiding the many minor variations in level which require the inclusion of additional stairs or ramps, leading to the restriction of easy circulation and wasted office space. The increasing use of computer technology and data-link services requires extensive cabling which is best accommodated within total-access raised floors. The heat given off by computer equipment has partly influenced the current practice of including fully air-conditioned office environments. Again consistent floor levels facilitate the installation of a standard type and more energy-efficient air-conditioning equipment.

Considerable emphasis was placed by Scottish Provident Institution in its consultation document of 1987 on creating an expanded Head Office complex which reflects a particular image for the organisation. SPI's required statement of corporate identity led them to propose a building in the architectural vocabulary of their existing Head Office at No. 6 St Andrew Square.

Negotiations between SPI and the planners

SPI's objective in submitting their Head Office expansion project consultation document to the planners was: to encourage the local planning authority in conjunction with the Historic Buildings and Monuments Directorate for Scotland; to establish design guidelines for the project, taking into account the owner's operational requirements and in particular their need for continuity of structural floor levels established in the existing Head Office buildings; and for a corporate design image. The planners required SPI to demonstrate that the project was not being pursued on a speculative basis, and SPI responded by outlining their plans to take up the proposed office space which, on available staff projections, was to work towards 100 per cent occupation of the new buildings within a five-year period. The planners formulated their response to the SPI proposals following internal discussions with the Historic Buildings and Monuments Directorate for Scotland, and guideline recommendations were sent to SPI in August 1987. The guidelines pointed out that Nos 4 and 5 St Andrew Square were statutorily listed in 1974, both for their intrinsic architectural merit and also for the maintenance of the harmony of stone buildings in the square. SPI were reminded that the arguments against the demolition of the buildings were publicly rehearsed in 1975 when similar proposals had been the subject of a listed building consent application by SPI; on that occasion, application had been refused following strong objections by the major amenity organisations and the recommendations of the then Director of Planning at the local authority. However, the 1987 guidelines accepted that the J. A. Carfrae building at No. 5 has plain detailing and an austere interior lacking the quality of the Italianate No. 4, which if restored to its former glory complete with original Corinthian porches and other detailing would assume even greater distinction.

SPI were advised by the Deputy Director of Planning at the City of Edinburgh District Council, who wrote setting out the recommended guidelines, that he would not object to the replacement of No. 5 with a building of 'distinctive and outstanding architecture, which did not replicate No. 6, and had less glass'. The view was also expressed that the facade of No. 4 and its main interior possessed too great an architectural value to lose, although some tidying up at the rear might be possible. The suggestion was made to SPI that the front section of No. 4 would make good boardrooms. The guidelines also suggested the acquisition of other properties in the immediate vicinity of SPI's headquarters, but these were dismissed as being operationally unpractical and financially non-viable.

SPI agreed to investigate the possibility of redeveloping No. 5 in its entirety and the part redevelopment of No. 4, a technique which had been used by another financial institution, TSB, at its headquarters nearby in George Street, where the facade and front rooms were retained or reinstated. By May 1988, a preliminary design scheme had been developed following regular consultations with the local planning authority, comprising the retention of the facade and principal rooms of No. 4 with redevelopment to the rear and the total replacement of No. 5 with a new stone-fronted building. The preliminary scheme was presented to the Royal Fine Art Commission for Scotland in May 1988. In response, the Commission first expressed their pleasure at being able to see a proposed scheme at an early stage before developers and designers become too committed to a particular idea, presumably in the knowledge that the Commission's views in this case were more likely to be taken into account. In terms of the proposed development, the Commission strongly expressed the feeling that No. 5 St Andrew Square should not be sacrificed. In their view 'this handsome Edwardian building' contributed to the remarkably successful composition of the southern side of the square, although it was felt this could be improved by the reinstatement of the Corinthian porches and other detailing of No. 4 and the eventual replacement of No. 8 St Andrew Square, on the corner of South St David Street. The Commissioners were disturbed that those concerned with guiding the design of the development were too readily discussing the total removal of a listed building. However, it was indicated that the Commissioners would not be averse to sensitive redevelopment behind the facade of No. 5 where the internal spaces were related to the fenestration of the facade. The contemplation of the removal of the Edwardian building would depend on its replacement by a building which in 30 years' time would be likely to be thought worthy of listing. It was recognised that Nos 4 and 6 St Andrew Square, are each in their own way distinguished buildings of the nineteenth and twentieth centuries respectively, and that the developers' objectives were to give due prominence to No. 4 and to enhance SPI's corporate identity by providing an appropriate setting for the frontage of No. 6. The view was reported that these objectives could best be realised by a contemporary sensitive, distinctive building which would not vie with its very different neighbours.

While it was felt that this might be achieved by a skilled modern designer the Commissioners doubted that a modern design could be more successful for the site at No. 5 than the existing Edwardian building.

Turning to the proposed design for a new building at No. 5 St Andrew Square, the Commissioners considered it most inferior to the existing building and the adopted approach demonstrated 'how unwise it is to try to refer simultaneously to a good 1883 building and a good 1961 building in one and the same facade'. The Commissioners could find nothing in the design worth salvaging and recommended a fresh start if a new building was the favoured option, and recommended that an architect who had specialised experience in designing a suitable infill frontage ought to be part of the development team.

The planners responded by largely agreeing with the advice of the Royal Fine Art Commission apart from the idea of building behind the facade of No. 5. This is interesting, given the outcome in this case. The planners were promoting the creation of an outstanding example of contemporary architecture to replace No. 5 and to this effect were, as well as the Royal Fine Art Commission, suggesting the appointment of an appropriate consultant architect to work on the project. Consequently, a shortlist was drawn up and approved by the local planning authority, from which Roland Wedgwood Associates were appointed to the development team. From July 1988 the consultant architects undertook numerous consultations with the Royal Fine Art Commission and, with their encouragement, developed a design involving the replacement of both Nos 4 and 5 St Andrew Square by a building featuring a curved curtain-wall facade with two giant stone portals standing in front of it, the narrow space between them representing the mid-point of the southern side of St Andrew Square. In September 1988 the Commissioners expressed the view that this building 'could be of the right scale and character in the post-1961 context, and better in these respects than the existing listed buildings'.

Planning and listed building consent applications are submitted

Roland Wedgwood continued to refine the design of the proposed building to replace Nos 4 and 5 St Andrew Square, following which planning and listed building consent applications were submitted to the local authority in November 1988. Having being consulted, the Royal Fine Art Commission reiterated their enthusiasm for the refined design. It was felt that the horizontal divisions of the curtain wall and the vertical slit windows in the stone walls of the porticos, which were formerly portals, would improve the proportions of the front and the curved roof would enhance the three-dimensional quality of the building.

Numerous letters of representation were received from amenity societies and private individuals following the submission of the applications. One of the most influential amenity groups, the Cockburn Association, was very supportive of the proposals considering, on balance, that the demolition of the listed buildings at Nos 4 and 5 could be justified because of the outstanding design of Roland Wedgwood's replacement building. Significantly, the secretary of the Cockburn Association wrote to the local planning authority dissenting from the views of his association, drawing attention to the architectural qualities of No. 4 and suggesting that there would be no difficulty in finding new owners or tenants for Nos 4 and 5. He also maintained that to allow demolition of two listed buildings would set an unfortunate precedent and encourage other owners of listed buildings to seek demolition. The Scottish Civic Trust accepted that deeper thought had gone into the preparation of the scheme than when the 1976 proposals had been formulated, but did not consider that the demolition of No. 4 St Andrew Square, would be justified. It was felt that the western half of the site could be redeveloped with a modern frontage of high quality which would have the advantage of preserving the memory of the original feu pattern. The Architectural Heritage Society of Scotland revised its opinions following a presentation by SPI's professional team. Its initial view was that the proposed building would be an inadequate replacement for the listed buildings. The Society's subsequent view was that the proposed design possessed quality, but it was feared that demolition of the listed buildings would damage the integrity of the Historic Buildings and Monuments Directorate lists of protected buildings, and an objection to the proposals was maintained.

Roland Wedgwood advised the planning authority that the demolition of Nos 4 and 5 was considered when it was found that to retain No. 4 would involve a great deal of reinstatement of detailing because the original stone used was of such poor quality. Also, both No. 4 (1883) and No. 5 (1903) had replaced the original buildings of 1770 and have now been in existence for nearly as long as the original buildings were. Further, the McLachlan building at No. 4 related poorly to the scale of the rest of the southern side of St Andrew Square. The point was made by Roland Wedgwood that the design for the proposed replacement building had evolved following regular meetings with both the Royal Fine Art Commission for Scotland and the city planners.

An important part of the brief had been to provide the maximum possible floor space for SPI with consistent floor levels throughout the complex. The design concept established a centre line and north–south axis of St Andrew Square, which was achieved by the curved facade centred on the axis and the symmetrical arrangement of the pylons. The design aimed at creating an awareness of sun on the north elevation, assisted by the glazed edge to the roof and curved section to the plant rooms. Also there was a historical reference to the original two feus emphasised by the pylons and the lack of a central entrance. In an attempt to contextualise the new facade with both the historic elements and the 1960s building on the southern side of the square, the substantial area of stone in the new facade echoed the general solid-to-void proportions of the other buildings and the new facade responded to the different planes of No. 6. Roland Wedgwood described the design as being 'a combination of the fully glazed building as wanted by the client, albeit set further back than he wished, and the stone facade which is what we consider to provide the necessary "dress" for a new building in this square. Much of the excitement comes from the inter-play between these two elements'.

Given that the development proposals had been formulated in consultation with the planners and the Royal Fine Art Commission, the Director of Planning of the City of Edinburgh District Council presented a favourable report to the Planning and Development Committee, recommending approval of the application for planning permission and listed building consent. The report recognised that opposition to the proposals expressed by some organisations and individuals were based on the fear that allowing the demolition of two statutorily listed buildings would undermine conservation policies within the central area of Edinburgh and encourage redevelopment in less justified circumstances. However, it was argued that SPI's development at No. 6 St Andrew Square in the 1960s was the last significant central area office development of post-war architecture and, in the 20 years since its construction, the city centre had very little to exhibit in the way of architectural progress. The committee were reminded that the consultant architect for the scheme was well known for his thoughtful approach to design and detail and had, in this case, produced an able solution to the problem of relating a new facade to the architecture of St Andrew Square. In view of the importance of the detailing to the successful completion of the building it was considered essential for the consultant architect to supervise construction through to completion. Finally, the report expressed the view that approving the scheme would not create any precedents in terms of relaxing conservation policies and, if the development was successfully executed in detail, should encourage architects generally to avoid the pastiche. The matter was considered by a Planning Sub-Committee, at which members wished to visit the site before making a recommendation to the full Planning and Development Committee. Following the site visit, the Sub-Committee recommended refusal of the applications on the grounds that the proposals would result in the loss of two Category B statutorily listed buildings, which would adversely affect the character of the New Town Conservation Area.

When this recommendation was announced the Royal Fine Art Commission wrote to the Committee Chair expressing concern and re-emphasising the Commission's support for the proposals. The Commission argued that: rejection of the scheme would be a serious misfortune for Edinburgh; the courageous architectural concept was better in scale and character than the existing buildings at Nos 4 and 5; that the proposed building would be elegant, original and striking without being aggressive or overwhelming; and that Edinburgh should be seen to be progressive as well as conservationist, which meant that there should be a place for good contemporary architecture. It was also argued that alternative designs would most likely be dull and commonplace and if the present design proved to be of insufficient merit for the District Council, existing buildings should remain standing.

Both Roland Wedgwood and SPI wrote to the District Council to try to overturn the recommendation. Roland Wedgwood highlighted the ideas which underpinned his approach to the redevelopment. An important starting point in his thinking was the questioning of the validity of preserving the forward part of No. 4, given the extensive external refurbishment required and total reconstruction of the interior using materials which would give adequate fire protection. He urged the local authority to be 'Bloody Bold and Resolute' in letting the development go ahead. SPI stressed the importance of the development to their successful expansion plans and outlined their approach to evolving an appropriate scheme for the site, which they felt had been acceptable to the planners and had received the support of the Royal Fine Art Commission for Scotland. In consequence, they urged the planning authority to re-examine the proposals and to allow the construction of a building which would make an outstanding contribution to the architecture of St Andrew Square. When the full Planning Development Committee met on 15 March 1989 the recommendation of the Sub-Committee was upheld and permission for the proposals refused. My research reveals that two particular actors were particularly influential at the Planning Committee stage of the decision-making process. First, the fact that the Chair of the Cockburn Association had written dissenting from the views expressed by the amenity group carried significant weight. Second, the Chair of the Planning Committee favoured the retention of the facades of Nos 4 and 5 St Andrew Square. Traditionally, the position of Chair in the local government committee system is one of power and influence in terms of decision-making.

SPI subsequently appealed against the decision of the planning authority to the Secretary of State for the Environment but to no avail, the appeal being dismissed. They then turned to facadism as a solution to the problem, which has led to the present outcome.

Summary

This case primarily demonstrates that facadism provides a welcome escape route for decision-makers who still fear the consequences of replacing albeit ordinary examples of historic buildings with contemporary architecture. The lack of faith in contemporary architecture that exists – which is often a driving force behind facadism being an outcome – is emphasised all the more in this case by the weight of informed opinion that suggested that Roland Wedgwood's design for a new building was of outstanding quality.

Planners, developer and consultant architect had worked in close harmony to evolve a design which, in terms of a wide cross-section of professional opinion, would have considerably enhanced the townscape and provided a valuable continuity of architectural tradition in the square. With this in mind, we see that facadism is not a development approach devised in a technical expertise vacuum, but is very much subject to the power relations within the planning decision-making process. In other words, in some cases facadism may not be the professionals' chosen option but rather a means of satisfying public or elected member opinion.

Finally, in this case, facade retention provided the only realistic option for SPI in terms of expanding their head office in a manner satisfactory to them and the Planning Committee. It allowed them far greater scope to create sufficient acceptable office space than converting the existing buildings would have done, and so facadism may be seen here as a pragmatic response to a conflict of interests.

NO AFTERWORD

There are no further conclusions I would wish to draw at the end of this book. I have offered many thoughts on facadism from a range of perspectives, and summaries of the facts have been given relating to the case studies in the last chapter. Now I feel it is appropriate to leave readers with their own thoughts and conclusions in the hope that the book has for them been both stimulating and enlightening.

NOTES AND REFERENCES

1 INTRODUCTION

1 A. Dobby (1978) *Conservation and Planning.*

2 Heritage planning is a process which has emerged out of developments in three overlapping facets of planning expertise and activity. The first facet involves the growth of a concern to preserve and present buildings, relics and place-associations of the past and a consequent re-evaluation of the worth of built environments of the past. Secondly, a contemporary industry has emerged which utilises historic resources to support modern urban functions and meet current demands, often through heritage tourism. The third facet concerns the use of conserved built environments in area planning and management initiatives to regenerate and renew local economies and communities. G. J. Ashworth (1991) in *Heritage Planning*, traces the evolution of the synthesis of these facets into the process of heritage planning. In some cases, facadism may be seen to have a significant role to play within such a process.

3 Quoted in J. Glancey (1989) *New British Architecture.*

4 Open-ended discussions were considered the most appropriate technique for describing and analysing the complex interaction of interests involved in the decision-making processes that lead to development outcomes; see R. K. Yin (1982) 'Studying the Implementation of Public Programmes' in W. Williams (1982) *Studying Implementation.* Structured interviews are notoriously inefficient devices for acquiring explanations of situations, see M. Fullan and A. Pomfret (1977) 'Research on Curriculum and Instruction Implementation' in *Review of Educational Research* 47: 335–97.

2 WHAT IS FACADISM?

1 View expressed by D. T. Martin, City of Glasgow Planning Department, in a letter dated 25 November 1991.

2 Expressed in reply to author's questionnaire, 1991.

3 G. J. Ashworth (1991) in *Heritage Planning*, p. 140.

4 John Outram quoted in J. Glancey (1989) p. 188.

5 Ashworth (1991).

6 View expressed by Dr G. J. Ashworth, University of Groningen, in letter to author dated 27 October 1992.

3 ARCHITECTURAL PERSPECTIVES

Pre-modern

1 A. J. Brook, Bristol City Council Conservation Officer, writing in *Avon Conservation News* (1986), Avon County Council, August.

2 This view, expressed by SAVE, was included in an unpublished study by D. Highfield, York University, 1982.

3 See P. Zucker, *Town and Square* (1959), p. 177.

4 For example, the opinion of A. J. Scrase, Lecturer in Town and Country Planning, Bristol Polytechnic, during the author's Bachelor of Town Planning Degree course (1985–88), the work of whom is gratefully acknowledged.

5 See R. Scruton (1979) *The Aesthetics of Architecture*. This book is of fundamental significance in terms of architectural perspectives of concepts such as facadism.

6 See note 2 above.

7 A. W. Pugin (1836) *Contrasts; or a Parallel between the Noble Edifices of the Fourteenth and Fifteenth Centuries and Similar Buildings of the Present Day. Showing the Decay of Taste*, first published by the author. A finely illustrated work in which Pugin compared the great medieval buildings with those of later years. However, his choice of comparisons was often highly prejudiced, with the medieval examples being of great quality and later buildings of more modest standard. Pugin's later works displayed more balanced, cogent views in which he combined his deep convictions with as thorough a comprehension of the structure, design and decoration of Gothic architecture as had rarely before been achieved.

8 See D. Yarwood (1985) *Encyclopaedia of Architecture*, p. 287.

9 See Scruton (1979).

10 J. Glancey (1989) *New British Architecture.*

Modern

11 J. Glancey (1989) *New British Architecture*; see also Nikolaus Pevsner (1943), *An Outline of European Architecture.*

12 Survey of derelict listed buildings carried out by Bristol City Council, 1974.

Postmodern

13 Quoted from Robert Venturi's *Complexity and Contradiction in Architecture* (1966), in which he challenged Modernists' condemnation of redundant space superfluous to functional requirements and in particular the views of Mies van der Rohe, whose *Existenzminimum* housing designed with Walter Gropius, where factory worker housing was provided with ceilings of no more than seven feet in height, provided a typical model for the Modern movement.

14 William Empson, *Seven Types of Ambiguity* (1955), p. 174.

4 TOWNSCAPE PERSPECTIVES

1 See K. Lynch (1960), p. 9.
2 That is, 'imageability' according to Lynch (1960).
3 Jerry DeMuth, in an article in *Estates Gazette*, 16 July 1988.
4 ibid.
5 ibid.
6 See G. J. Ashworth, *Heritage Planning* (1991).

5 SOCIAL AND ECONOMIC PERSPECTIVES

1 D. Eversley (1974) 'Conservation for the Minority?', *Built Environment* 3 (1): 14.
2 The shopping developments referred to are as follows: Bath – The Colonnades, Bath St; Cheltenham – Regent Arcade, High St, Regent St; Exeter – Marks & Spencer's, High St, Queen St, Goldsmith St; Barnstaple – Green Lanes, High St, Joy St.
3 Those readers wishing to investigate the process of gentrification further may refer to N. Smith and P. Williams (eds) (1986) *Gentrification of the City*.
4 For example, historic preservation in the USA was significantly bolstered between 1981 and 1986 by the availability of tax credits and accelerated depreciation benefits for investments in historic property. See P. H. Gleye (1988) 'With Heritage so Fragile: A Critique of the Tax Credit Program for Historic Building Rehabilitation', *Journal of the American Planning Association* 54: 482–8.
5 Norwich City Council indicated in 1991 their difficulties experienced in trying to save unlisted but interesting historic buildings in their entirety because of VAT regulations which in some circumstances allowed developers to retain one wall of the original building and receive tax benefits on the rest of the reconstruction. The complexity of the VAT regulations and their impact on development decisions is a subject in its own right, and no further analysis is offered here. Readers wishing to research the subject further are advised to approach HM Customs and Excise.

6 FACADISM AND URBAN CONSERVATION

1 S. Owen (1976) *Planning Outlook*, vol. 18, Spring: 41.
2 F. Tibbalds (1988) 'Mind the Gap', *The Planner*, March.
3 S. Gleave and D. Wheeler (1988) results of research and article 'Development Framework Planning: South Fulham Riverside', *The Planner*, March: 26.
4 See G. J. Ashworth (1991).
5 See article 'Vital Steps to Becoming Greener' in *Voluntary Housing*, August 1992, pp. 9–11.
6 Glastonbury and Street Area Local Plan, deposited document, July 1991, published by Mendip District Council, Somerset, pp. 5, 6.
7 Town and Country Planning (Amendment) Act 1972.
8 Portland Square Conservation Area policy document (1974), published by Bristol City Council.
9 View expressed by English Heritage in letter to author dated 7 October 1987.

10 *Chester Conservation Review Study Draft Report* (1984), published by Chester City Council.
11 Indicated by author's own research; see pp. 122–6.
12 Architect D. Insall in a letter to the Department of the Environment, quoted in *Chester Conservation Review Study Draft Report* (1984).
13 John Honeyman's exquisitely crafted former warehouse is one of Glasgow's rare cast-iron-fronted City Centre buildings. Built in 1872, it became a ballroom and restaurant in the 1920s. Changing fashions led to the building's decline and a threat to its future. CIS Ltd, Manchester, carried out the development which saved the facades and interior decorative glass of the original building and added two more cast-iron bays to match those of Ca'D'Oro replacing an adjoining ferroconcrete building.
14 F. Tibbalds (1971) 'Planning for Conservation', *RIBA Journal*, August, p. 361.
15 ibid.
16 Owen (1976).
17 From unpublished study by D. Highfield (1982), University of York.
18 K. McGovern and B. Wola (1987) 'Quest for New Approaches to Town Centres', *Planning*, 27 November.
19 *Avon Conservation News* (1986), published by Avon County Council, issue no. 23, August.
20 Highfield (1982) – see note 17 above.
21 Jerry DeMuth (1988) 'American Architecture', *Estates Gazette*, 16 July.
22 P. F. Smith (1974) 'Familiarity Breeds Contentment', *The Planner*, vol. 60, November.
23 Roy Worskett, quoted in McGovern and Wola (1987).
24 T. Cantell (1975) 'Why Conserve?', *The Planner*, vol. 61, January.
25 P. F. Smith (1974).
26 Lord Clark quoted in Cantell (1975).
27 *Chester Conservation Review Study Draft Report* (1984).
28 View expressed by the Curator in Field Archaeology, City of Bristol Museum and Art Gallery, in a memorandum to the City Planning Officer, Bristol, dated 20 October 1987.

SELECT BIBLIOGRAPHY

Ashworth, G. J. (1991) *Heritage Planning*, Groningen, The Netherlands: Geo Pers.
Barrett, M. and Fudge, S. (eds) (1981) *Policy and Action: Essays on the Implementation of Public Policy*, London: Methuen.
Burke, G. (1976) *Townscapes*, Harmondsworth: Penguin.
Cadman, D. and Austin-Crowe, L. (1978) *Property Development*, London: Spon.
Cherry, G. (1972) *Urban Change and Planning*, Henley-on-Thames: Foulis.
Ching, F. D. K. (1989) *Architecture: Form, Space and Order*, New York: Van Nostrand Reinhold Co.
Cullen, G. (1961) *Townscape*, London: The Architectural Press.
—— (1961) *The Concise Townscape*, London: The Architectural Press.
Davies, C. (1975) *Permissive Britain: Social Change in the '60s and '70s*, London: Pitman.
Department of the Environment (1987) *Historic Buildings and Conservation Areas – Policy and Procedures*, Circular 8/87, London: Department of the Environment.
Dobby, A. (1978) *Conservation and Planning*, London: Hutchinson.
Empson, W. (1955) *Seven Types of Ambiguity*, New York: Meridian Books.
Glancey, J. (1989) *New British Architecture*, London: Thames & Hudson.
Godfrey, W. H. (1944) *Our Building Inheritance*, London: Faber & Faber.
Ham, C. and Hill, M. (1984) *The Policy Process in the Modern Capitalist State*, Brighton: Harvester Press.
Heckscher, A. (1962) *The Public Happiness*, New York: Atheneum.
Hewison, R. (1987) *The Heritage Industry*, London: Methuen.
Highfield, D. (1991) *Construction of New Buildings Behind Historic Facades*, London: Spon.
HRH The Prince of Wales (1989) *A Vision of Britain*, London: Doubleday.
Ittelson, W. H. (1960) *Visual Space Perception*, New York: Springer.
Jacobs, J. (1961) *The Death and Life of Great American Cities*, New York: Random House.
Jencks, C. (1970) *Post Modernism: The New Classicism in Art and Architecture*, London: Academy Editions.
—— (1973) *Modern Movements in Architecture*, Harmondsworth: Penguin.
—— (1977) *The Language of Post Modern Architecture*, London: Academy Editions.
Le Corbusier (1927) *Towards a New Architecture*, trans. by Frederich Etchells from *Vers une architecture*, London: The Architectural Press (first published in France by Editions Crès, 1923).

—— (1929) *The City of Tomorrow and its Planning*, trans. Frederich Etchells from *Urbanisme*, 8th edn, London: John Rohder; reprinted (1947) London: The Architectural Press.

—— (1954) *The Modulor*, London: Faber (first published in France as *Le Modulor*, 1948).

—— (1967) *The Radiant City*, London: Faber (first published in France as *La Ville radieuse*, 1933).

Lynch, K. (1960) *The Image of the City*, Cambridge, Mass.: MIT Press.

Pevsner, N. (1943) *An Outline of European Architecture*, Harmondsworth: Penguin.

—— (1968) *The Sources of Modern Architecture and Design*, London: Thames & Hudson.

Pocock, D. and Hudson, R. (1978) *Images of the Urban Environment – Focal Problems in Geography*, London: Macmillan.

Popper, K. (1957) *The Poverty of Historicism*, London: Routledge & Kegan Paul.

Pugin, A. W. (1969, reprint of 2nd edn text 1841; original 1836) *Contrasts; or a Parallel between the Noble Edifices of the Fourteenth and Fifteenth Centuries and Similar Buildings of the Present Day. Showing the Decay of Taste*, Leicester: Leicester University Press.

—— (1973, reprint of original 1841) *The True Principles of Pointed or Christian Architecture*, London: Academy Editions (first edition, 1841, published London: John Weale).

Rapoport, A. (1982) *The Meaning of the Built Environment*, Beverly Hills, Calif.: Sage.

Rapoport, A. and Kantor, A. (1967) 'Complexity and Ambiguity in Environmental Design', *AIP Journal*, 33 (4): 210–21.

Rasmussen, S. E. (1959) *Experiencing Architecture*, London: Chapman & Hall.

Richards, J. M. (1940) *An Introduction to Modern Architecture*, Harmondsworth: Penguin (revised editions in 1962 and 1970).

Ruskin, J. (1907) *The Seven Lamps of Architecture*, London: Dent (first published 1849; reprinted in 1963).

—— (1960) *The Stones of Venice*, vols I–III, London: Collins (first published 1851–53).

Scruton, R. (1979) *The Aesthetics of Architecture*, London: Methuen.

Sharp, T. (1968) *Town and Townscape*, London, Murray.

Sked, A. (1987) *Britain's Decline: Problems and Perspectives*, Oxford: Blackwell.

Smith, N. and Williams, P. (eds) (1986) *The Gentrification of the City*, Boston: Allen & Unwin.

Tugnutt, A. and Robertson, M. (1987) *Making Townscapes: A Contextual Approach to Building in an Urban Setting*, London: Mitchell.

Venturi, R. (1966) *Complexity and Contradiction in Architecture*, New York: Museum of Modern Art.

Vitruvius (1914) *The Ten Books on Architecture*, New York: Dover Publications, trans. from *De Architectura* (date of original publication unknown).

Watkin, D. (1977) *Morality and Architecture*, Oxford: Clarendon Press.

Williams, W. *et al.* (1982) *Studying Implementation*, New Jersey: Chatham House.

Worskett, R. (1969) *The Character of Towns*, London: The Architectural Press.

Yarwood, D. (1985) *Encyclopaedia of Architecture*, London: Batsford.

Zucker, P. (1959) *Town and Square*, Cambridge, Mass.: MIT Press.

INDEX

Page references in **bold** refer to plates

For Product Safety Concerns and Information please contact our EU
representative GPSR@taylorandfrancis.com
Taylor & Francis Verlag GmbH, Kaufingerstraße 24, 80331 München, Germany

www.ingramcontent.com/pod-product-compliance
Lightning Source LLC
LaVergne TN
LVHW080653011026
849160LV00014B/1408
9780415756020